D1586533

DATE REMOVED
- 2 JUL 2021
STOCK

advancing learning, changing lives

Decision
Mathematics 1

Edexcel AS and A-level
Modular Mathematics

Susie G Jameson

Contents

Longley Park Sixth Form College

ACC NO | 00026035

CLASS NO | LOAN TYPE

About this book

This book is designed to provide you with the best preparation possible for your Edexcel D1 unit examination:

- This is Edexcel's own course for the GCE specification.
- Written by Senior Examiners.
- The LiveText CD-ROM in the back of the book contains even more resources to support you through the unit.

Brief chapter overview and 'links' to underline the importance of mathematics: to the real world, to your study of further units and to your career

Finding your way around the book

Every few chapters, a review exercise helps you consolidate your learning

Detailed contents list shows which parts of the D1 specification are covered in each section

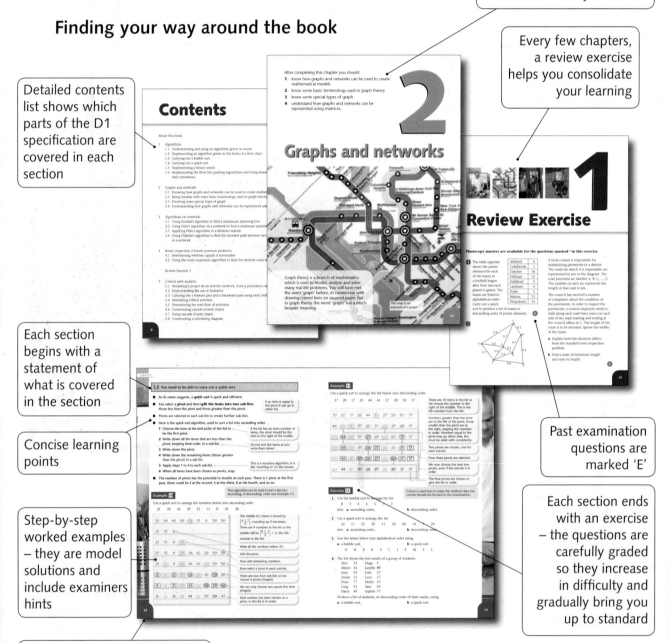

Each section begins with a statement of what is covered in the section

Concise learning points

Step-by-step worked examples – they are model solutions and include examiners hints

Past examination questions are marked 'E'

Each section ends with an exercise – the questions are carefully graded so they increase in difficulty and gradually bring you up to standard

Each chapter has a different colour scheme, to help you find the right chapter quickly

Each chapter ends with a mixed exercise and a summary of key points.

At the end of the book there is an examination-style paper.

LiveText software

The LiveText software gives you additional resources: Solutionbank and Exam café. Simply turn the pages of the electronic book to the page you need, and explore!

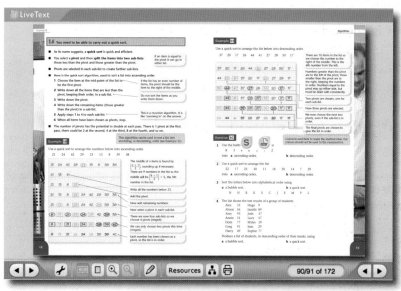

Unique Exam café feature:

- Relax and prepare – revision planner; hints and tips; common mistakes
- Refresh your memory – revision checklist; language of the examination; glossary
- Get the result! – fully worked examination-style paper with chief examiner's commentary

Solutionbank

- Hints and solutions to every question in the textbook
- Solutions and commentary for all review exercises and the practice examination paper

Published by Pearson Education Limited, a company incorporated in England and Wales, having its registered office at Edinburgh Gate, Harlow, Essex, CM20 2JE. Registered company number: 872828

Edexcel is a registered trademark of Edexcel Limited

Text other than chapter 5 © Susie Jameson 2009; chapter 5 © Peter Sherran 2009

12 11 10 09
10 9 8 7 6 5 4 3 2

British Library Cataloguing in Publication Data
A catalogue record for this book is available from the British Library on request.

ISBN 978 0 435519 19 3

Copyright notice
All rights reserved. No part of this publication may be reproduced in any form or by any means (including photocopying or storing it in any medium by electronic means and whether or not transiently or incidentally to some other use of this publication) without the written permission of the copyright owner, except in accordance with the provisions of the Copyright Designs and Patents Act 1988 or under the terms of a licence issued by the Copyright Licensing Agency, Saffron House, 6–10 Kirby Street, London EC1N 8TS (www.cla.co.uk). Applications for the copyright owner's written permission should be addressed to the publisher.

Edited by Susan Gardner
Typeset by Tech-Set Ltd
Illustrated by Tech-Set Ltd
Cover design by Christopher Howson
Picture research by Chrissie Martin
Cover photo/illustration © Edexcel
Index by Indexing Specialists (UK) Ltd
Printed in Italy by Rotolito

Acknowledgements
The author and publisher would like to thank the following individuals and organisations for permission to reproduce photographs:

iStockPhoto.com/Leigh Schindler p1; Alamy/David R. Frazier Photolibrary, Inc. p25; Pearson Education/Naki Photography p40; iStockPhoto.com/Ralph125 p63; Alamy/Ian Miles – Flashpoint Pictures p87; Shutterstock/ Graca Victoria p113; Rex Features/Micha Theiner p149

Every effort has been made to contact copyright holders of material reproduced in this book. Any omissions will be rectified in subsequent printings if notice is given to the publishers.

Disclaimer
This Edexcel publication offers high-quality support for the delivery of Edexcel qualifications.
Edexcel endorsement does not mean that this material is essential to achieve any Edexcel qualification, nor does it mean that this is the only suitable material available to support any Edexcel qualification. No endorsed material will be used verbatim in setting any Edexcel examination/assessment and any resource lists produced by Edexcel shall include this and other appropriate texts.
Copies of official specifications for all Edexcel qualifications may be found on the Edexcel website – www.edexcel.com.

After completing this chapter you should be able to:

1 use and understand an algorithm given in words

2 understand how flow charts can be used to describe algorithms

3 carry out a bubble sort, quick sort and binary sort

4 carry out the three bin packing algorithms and understand their strengths and weaknesses.

Algorithms

An **algorithm** is a precise set of instructions that is so clear that it will allow anyone, or even a computer, to use it to achieve a particular goal in a specified number of steps. Ideally, an algorithm should be written in such a way that it is easy to convert into a computer program.

There are strong links between the development of computer technology and the development of Decision Mathematics. Although we shall only be looking at small-scale examples, remember that most of the algorithms in this book have been developed to enable computers to solve large-scale problems.

A recipe is an example of an algorithm.

1.1 You need to be able to understand and use an algorithm given in words.

■ You have been using algorithms since you started school. Some examples of mathematical algorithms you have been taught are

- how to add several two-digit numbers
- how to multiply two two-digit numbers
- how to add, subtract, multiply or divide fractions.

■ It can be quite challenging to write a set of instructions that would enable a young child to do these tasks correctly.

Here are some more examples.

At the end of the street turn right and go straight over the cross roads, take the third left after the school, then ...

Affix base (B) to leg (A) using screw (F) and ...

Dice two large onions.
Slice 100 g mushrooms.
Grate 100 g cheese.

You will not have to write algorithms in the examination. You will need to implement them.

Example **1**

The 'happy' algorithm is

- write down any integer
- square its digits and find the sum of the squares
- continue with this number
- repeat until either the answer is 1 (in which case the number is 'happy') or until you get trapped in a cycle (in which case the number is 'not happy').

Show that

a 70 is happy

b 4 is unhappy.

a $7^2 + 0^2 = 49$	**b** $4^2 = 16$
$4^2 + 9^2 = 97$	$1^2 + 6^2 = 37$
$9^2 + 7^2 = 130$	$3^2 + 7^2 = 58$
$1^2 + 3^2 + 0^2 = 10$	$5^2 + 8^2 = 89$
$1^2 + 0^2 = 1$	$8^2 + 9^2 = 145$
so 70 is happy	$1^2 + 4^2 + 5^2 = 42$
	$4^2 + 2^2 = 20$
	$2^2 + 0^2 = 4$
	$4^2 = 16$
	so 4 is unhappy

Example 2

Implement this algorithm.

1 Let $n = 1$, $A = 1$, $B = 1$.

2 Write down A and B.

3 Let $C = A + B$.

4 Write down C.

5 Let $n = n + 1$, $A = B$, $B = C$. •————

6 If $n < 5$ go to 3.

7 If $n = 5$ stop.

> This instruction looks confusing.
> It simply means
> - replace n by $n + 1$ (add 1 to n)
> - A takes B's current value
> - B takes C's current value.

Use a trace table. •————

> A trace table is used to record the values of each variable as the algorithm is run.

Instruction step	n	A	B	C	Write down
1	1	1	1		
2					1, 1
3				2	
4					2
5	2	1	2		
6	Go to step 3				
3				3	
4					3
5	3	2	3		
6	Go to step 3				
3				5	
4					5
5	4	3	5		
6	Go to step 3				
3				8	
4					8
5	5	5	8		
6	Continue to step 7				
7	Stop				

> You may ask to complete a printed trace table in the examination. Just obey each instruction, in order.

> You may be asked what the algorithm does.

This algorithm produces the first few numbers in the Fibonnacci sequence.

Example 3

This algorithm multiplies the two numbers A and B.

1 Make a table with two columns.
   Write A in the top row of the left hand column and B in the top row of the right hand column.

2 In the next row of the table write:
   - in the left hand column, the number that is half A, ignoring remainders
   - in the right hand column the number that is double B.

> This famous algorithm is sometimes called 'the Russian peasant's algorithm' or 'the Egyptian multiplication algorithm'.

3 Repeat step 2 until you reach the row which has a 1 in the left hand column.

4 Delete all rows where the number in the left hand column is even.

5 Find the sum of the non-deleted numbers in the right hand column.
   This is the product AB.

Implement this algorithm when

a $A = 29$ and $B = 34$ **b** $A = 66$ and $B = 56$.

a

A	B
29	34
~~14~~	~~68~~
7	136
3	272
1	544
Total	986

So 29 × 34 = 986

> Each time the number in the left hand column is halved and the number in the right hand column is doubled.

> Step 4 means that rows where the number in the left hand column is even must be deleted before summing the right hand column.

b

A	B
~~66~~	~~56~~
33	112
~~16~~	~~224~~
~~8~~	~~448~~
~~4~~	~~896~~
~~2~~	~~1792~~
1	3584
Total	3696

So 66 × 56 = 3696

> Step 4 means that all rows where the number in the left hand column is even must be deleted before summing the right hand column.

Exercise 1A

1 a

> **1** Write the fractions in the form $\frac{a}{b}$ and $\frac{c}{d}$.
>
> **2** Let $e = ad$.
>
> **3** Let $f = bc$.
>
> **4** Print 'answer is $\frac{e}{f}$'.

Implement this algorithm with the fractions

i $2\frac{1}{4}$ **ii** $1\frac{1}{3}$.

b What does this algorithm do?

2 a Implement this algorithm.

1 Let $A = 1$, $n = 1$.   **2** Print A.   **3** $A = A + 2n + 1$.

4 Let $n = n + 1$.   **5** If $n \leqslant 10$ go to 2.   **6** Stop.

b What does this algorithm produce?

3

> **1** Input A, r.   **5** Let $r = s$.
>
> **2** Let $C = \frac{A}{r}$ to 3 decimal places.   **6** Go to 2.
>
> **3** If $|r - C| \leqslant 10^{-2}$ go to 7.   **7** Print r.
>
> **4** Let $s = \frac{1}{2}(r + C)$ to 3 decimal places.   **8** Stop.

> This algorithm requires you to use the modulus function. If $x \neq y$, $|x - y|$ is the positive difference between x and y. For example $|3.2 - 7| = 3.8$.

a Use a trace table to implement the algorithm above when

i $A = 253$ and $r = 12$,   **ii** $A = 79$ and $r = 10$,   **iii** $A = 4275$ and $r = 50$.

b What does the algorithm produce?

4 Use the algorithm in Example 3 to evaluate

a 244×125 **b** 125×244 **c** 256×123.

1.2 **You need to be able to implement an algorithm given in the form of a flow chart.**

- Flow charts are often used to design computer programs.

- There are three shapes of boxes which are used in the examination.

Start/End   Instruction   Decision

- The boxes in a flow chart are linked by arrowed lines.

- As with an algorithm written in words, you need to follow each step in order.

Example **4**

Box 1   Start

Box 2   Let $n = 0$

Box 3   Let $n = n + 1$ n is acting as a counter. It ensures that we stop after 10 terms.

Box 4   Let $E = 2n$

Box 5   Print E

Box 6   Is $n \geqslant 10$?   No   A decision box will contain a question to which the answer is either 'yes' or 'no'.

      Yes

Box 7   Stop

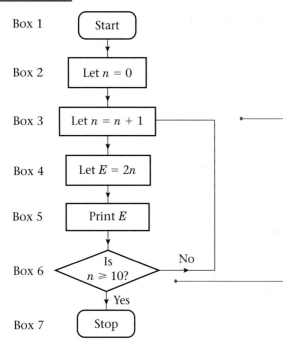

a Implement this algorithm using a trace table.

b Alter box 4 to read 'Let $E = 3n$' and implement the algorithm again.

How does this alter the algorithm?

a

n	E	box 6
0		
1	2	no
2	4	no
3	6	no
4	8	no
5	10	no
6	12	no
7	14	no
8	16	no
9	18	no
10	20	yes

Output is 2, 4, 6, 8, 10, 12, 14, 16, 18, 20

b

n	E	box 6
0		
1	3	no
2	6	no
3	9	no
4	12	no
5	15	no
6	18	no
7	21	no
8	24	no
9	27	no
10	30	yes

Output is 3, 6, 9, 12, 15, 18, 21, 24, 27, 30

This gives the first ten multiples of 3 rather than the first ten multiples of 2.

In a trace table each step must be made clear.

Example 5

This flow chart can be used to find the roots of an equation of the form $ax^2 + bx + c = 0$.

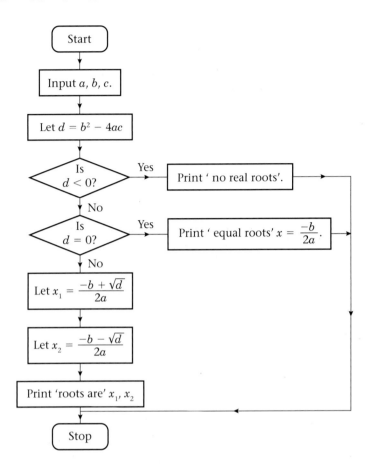

Start

Input a, b, c.

Let $d = b^2 - 4ac$

Is $d < 0$? — Yes → Print 'no real roots'.

No

Is $d = 0$? — Yes → Print 'equal roots' $x = \dfrac{-b}{2a}$.

No

Let $x_1 = \dfrac{-b + \sqrt{d}}{2a}$

Let $x_2 = \dfrac{-b - \sqrt{d}}{2a}$

Print 'roots are' x_1, x_2

Stop

Demonstrate this algorithm for these equations.

a $6x^2 - 5x - 11 = 0$ **b** $x^2 - 6x + 9 = 0$ **c** $4x^2 + 3x + 8 = 0$

a

a	b	c	d	$d < 0$?	$d = 0$?	x_1	x_2
6	-5	-11	289	no	no	$\frac{11}{6}$	-1

Roots are $x = \frac{11}{6}$ and $x = -1$

b

a	b	c	d	$d < 0$?	$d = 0$?	x_1	x_2
1	-6	9	0	no	yes	3	3

Equal roots are $x = 3$.

c

a	b	c	d	$d < 0$?
4	3	8	-119	yes

No real roots.

Example 6

Apply the algorithm shown by the flow chart on the right to the data
$u_1 = 10$, $u_2 = 15$, $u_3 = 9$, $u_4 = 7$, $u_5 = 11$.

What does the algorithm do?

> This is quite complicated because it has questions and a list of data. Tackle one step at a time.

	n	A	T	$T < A$?	$n < 5$?
box 1	1	10			
box 2	2				
box 3			15		
box 4				No	
box 6					Yes
box 2	3				
box 3			9		
box 4				Yes	
box 5		9			
box 6					Yes
box 2	4				
box 3			7		
box 4				Yes	
box 5		7			
box 6					Yes
box 2	5				
box 3			11		
box 4				No	
box 6					No
box 7	Output is 7				

The algorithm selects the smallest number from a list.

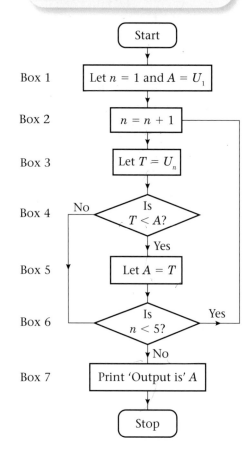

> The box numbers have been included to help you to follow the algorithm. You do not need to include them in the examination.

Exercise 1B

1 Apply the flow chart in Example 5 to the following equations.

    a $4x^2 - 12x + 9 = 0$ **b** $-6x^2 + 13x + 5 = 0$ **c** $3x^2 - 8x + 11 = 0$

2 **a** Apply the flow chart in Example 6 to the following data.

    i $u_1 = 28$, $u_2 = 26$, $u_3 = 23$, $u_4 = 25$, $u_5 = 21$

    ii $u_1 = 11$, $u_2 = 8$, $u_3 = 9$, $u_4 = 8$, $u_5 = 5$

b If box 4 is altered to $\langle$ Is $T > A?$ $\rangle$, how will this affect the output?

c Which box would need to be altered if the algorithm had to be applied to a list of 8 numbers?

3 Euclid's algorithm is applied to two non-zero integers a and b.

  a Apply Euclid's algorithm to

    i 507, 52

    ii 884, 85

    iii 4845, 3795

  b What does the algorithm do?

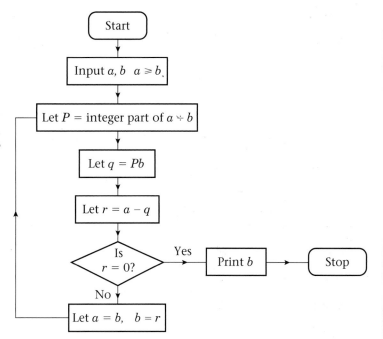

4 The equation $2x^3 + x^2 - 15 = 0$ may be solved by the iteration

$$x_{n+1} = \sqrt[3]{\frac{15 - x^2}{2}}$$

using the chart opposite.

  a Use $a = 2$ to find a root of the equation.

  b Use $a = 20$ to find a root of the equation. What do you notice?

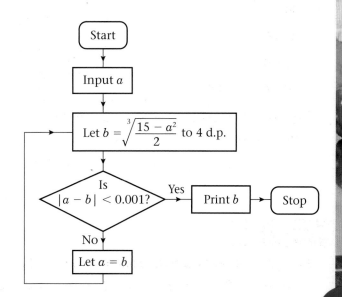

1.3 You need to be able to carry out a bubble sort.

■ A common data processing task is sorting an unordered list into alphabetical or numerical order.

■ In a **bubble sort** we compare **adjacent items**.

■ This is the bubble sort algorithm.

1 Start at the beginning of the list. Pass through the list and compare adjacent values. For each pair of values

- if they are in order, leave them
- if they are not in order, swap them.

2 When you get to the end of the list, repeat step **1**.

3 When a pass is completed without any swaps, the list is in order.

> There are many sorting algorithms. Only two need to be learnt for the examination – the bubble sort and the quick sort (see section 1.4).

> The elements in the list 'bubble' to the end of the list in the same way as bubbles in a glass of fizzy drink rise to the top of the glass. This is how the algorithm got its name.

Example 7

Use a bubble sort to arrange this list into ascending order.

24   18   37   11   15   30

(24 18) 37 11 15 30	1st comparison: swap
18 (24 37) 11 15 30	2nd comparison: leave
18 24 (37 11) 15 30	3rd comparison: swap
18 24 11 (37 15) 30	4th comparison: swap
18 24 11 15 (37 30)	5th comparison: swap
18 24 11 15 30 37	End of first pass

After the second pass the list becomes

18 11 15 24 30 37

After the third pass the list is

11 15 18 24 30 37

The fourth pass produces no swaps, so the list is in order.

> In the examination you may be asked to show each comparison for one pass, but generally you will only be required to give the state of the list after each pass.

> We now return to the start of the list for the second pass.

> Think of it as a race. The faster moving cars overtake the slower ones.

Example 8

Use a bubble sort to arrange these letters into alphabetical order.

A   L   G   O   R   I   T   H   M

After 1st pass	A G L O I R H M T
After 2nd pass	A G L I O H M R T
After 3rd pass	A G I L H M O R T
After 4th pass	A G I H L M O R T
After 5th pass	A G H I L M O R T
After 6th pass	A G H I L M O R T

After one pass at least the last letter is in its correct place.

After the second pass at least the end two letters are in place, and so on.

No swaps, so list is in order.

Be careful not to 'lose' any items during the sort!

Example 9

Use a bubble sort to arrange these numbers into descending order.

39   37   72   39   17   24   48

(39 57) 72 39 17 24 48	39 < 57 so swap
57 (39 72) 39 17 24 48	39 < 72 so swap
57 72 (39 39) 17 24 48	39 ≮ 39 so leave
57 72 39 (39 17) 24 48	39 ≮ 17 so leave
57 72 39 39 (17 24) 48	17 < 24 so swap
57 72 39 39 24 (17 48)	17 < 48 so swap
57 72 39 39 24 48 17	
After 1st pass:	57 72 39 39 24 48 17
After 2nd pass:	72 57 39 39 48 24 17
After 3rd pass:	72 57 39 48 39 24 17
After 4th pass:	72 57 48 39 39 24 17
No swaps in next pass, so list is in order.	

Note that the 48 is now in between the two 39s. Do not treat the two 39s as one term.

11

1.4 You need to be able to carry out a quick sort.

■ As its name suggests, a **quick sort** is quick and efficient.

■ You select a **pivot** and then **split the items into two sub-lists**: those less than the pivot and those greater than the pivot.

> If an item is equal to the pivot it can go in either list.

■ Pivots are selected in each sub-list to create further sub-lists.

■ Here is the quick sort algorithm, used to sort a list into ascending order.

1 Choose the item at the mid-point of the list to be the first pivot.

> If the list has an even number of items, the pivot should be the item to the right of the middle.

2 Write down all the items that are less than the pivot, keeping their order, in a sub-list.

> Do not sort the items as you write them down.

3 Write down the pivot.

4 Write down the remaining items (those greater than the pivot) in a sub-list.

5 Apply steps 1 to 4 to each sub-list.

> This is a recursive algorithm. It is like 'zooming in' on the answer.

6 When all items have been chosen as pivots, stop.

■ The number of pivots has the potential to double at each pass. There is 1 pivot at the first pass, there could be 2 at the second, 4 at the third, 8 at the fourth, and so on.

Example 10

> This algorithm can be used to sort a list into ascending, or descending, order (see Example 11).

Use a quick sort to arrange the numbers below into ascending order.

21   24   42   29   23   13   8   39   38

21	24	42	29	(23)	13	8	39	38
21	13	8						
21	13	8	[23]					
21	13	8	[23]	24	42	29	39	38
21	(13)	8	[23]	24	42	(29)	39	38
(8)	[13]	(21)	[23]	(24)	[29]	42	(39)	38
[8]	[13]	[21]	[23]	[24]	[29]	(38)	[39]	(42)
8	13	21	23	24	29	38	39	42

The middle of n items is found by $\left[\dfrac{n+1}{2}\right]$, rounding up if necessary. There are 9 numbers in the list so the middle will be $\left[\dfrac{9+1}{2}\right] = 5$, the 5th number in the list.

Write all the numbers below 23.

Add the pivot.

Now add remaining numbers.

Now select a pivot in each sub-list.

There are now four sub-lists so we choose 4 pivots (ringed).

We can only choose two pivots this time (ringed).

Each number has been chosen as a pivot, so the list is in order.

Example 11

Use a quick sort to arrange the list below into descending order.

37   20   17   26   44   41   27   28   50   17

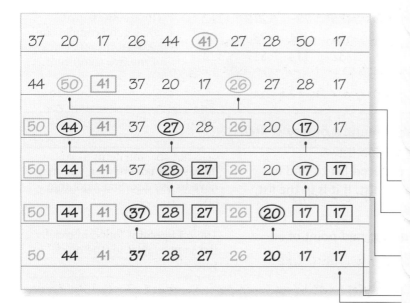

There are 10 items in the list so we choose the number to the right of the middle. This is the 6th number from the left.

Numbers greater than the pivot are to the left of the pivot, those smaller than the pivot are to the right, keeping the numbers in order. Numbers equal to the pivot may go either side, but must be dealt with consistently.

Two pivots are chosen, one for each sub-list.

Now three pivots are selected.

We now choose the next two pivots, even if the sub-list is in order.

The final pivots are chosen to give the list in order.

Exercise 1C

Colour is used here to make the method clear, but colours should *not* be used in the examination.

1 Use the bubble sort to arrange the list

8   3   4   6   5   7   2

into   **a** ascending order,   **b** descending order.

2 Use a quick sort to arrange the list

22   17   25   30   11   18   20   14   7   29

into   **a** ascending order,   **b** descending order.

3 Sort the letters below into alphabetical order using

a a bubble sort,   **b** a quick sort.

N   H   R   K   S   C   J   E   M   P   L

4 The list shows the test results of a group of students.

Alex   33   Hugo   9
Alison   56   Janelle 89
Amy   93   Josh   37
Annie   51   Lucy   57
Dom   77   Myles   19
Greg   91   Sam   29
Harry   49   Sophie 77

Produce a list of students, in descending order of their marks, using

a a bubble sort,   **b** a quick sort.

5 This is a question to do using cards, *not* by writing things down!
Sort the numbers listed below into ascending order using

> Write these 50 numbers on 50 cards and sort them by swapping cards.

a a bubble sort, **b** a quick sort.

453	330	405	792	516	162	465	870	431
927	129	348	34	107	64	253	382	411
147	389	597	414	620	425	73	275	212
482	302	52	868	144	65	471	930	766
243	578	274	630	281	732	114	517	322
748	517	492	331					

> This question gives you a good comparison between the efficiencies of the two algorithms.

1.5 You need to be able to implement a binary search.

■ A binary search will search an **ordered** list to find out whether a particular item is in the list. If it is in the list, it will locate its position in the list.

> If the list is *not* in order, you may have to use a sorting algorithm first.

■ A binary search concentrates on the mid-point of an ever-halving list, so it is very quick.

> The mid-point is found in exactly the same way as for the quick sort.

■ Here is the binary search algorithm.

To search an ordered list of n items for a target T:

1 Select the middle item in the list, m $\left(\text{use } \dfrac{n+1}{2} \text{ and round up if necessary}\right)$,

2 i if $T = m$, the target is located and the search is complete,

   ii if T is before m, it cannot be in the second half of the list, so that half, and m, are discarded,

   iii if T is after m, it cannot be in the first half of the list, so that half, and m, are discarded.

> If $T \neq m$, the pivot, m, and half the list are discarded. In each pass the list length halves.

3 Repeat steps **1** and **2** to the remaining list until T is found. (If T is not found it is not in the list.)

Example 12

Use the binary search algorithm to try to locate

a Robinson, **b** Davies

in the list below.

1 Bennett
2 Blackstock
3 Brown
4 Ebenezer
5 Fowler
6 Laing
7 Leung
8 Robinson
9 Saludo
10 Scadding

> Remember that a search can be unsuccessful. You may be asked to try to locate something that is not in the list. You must *demonstrate* that it is not in the list.

> Imagine the names are in order, but sealed in numbered envelopes. Each time you choose a pivot it is like opening that envelope.

[a] is used to mean the smallest integer $\geqslant a$.

a The middle name is $\left[\dfrac{10 + 1}{2}\right] = [5.5] = 6$ Laing.

Robinson is after Laing, so the list reduces to

7 Leung

8 Robinson

9 Saludo

10 Scadding.

Since Robinson is after Laing, Robinson cannot be in the first part of the list. So we delete the pivot, 6, and the first part of the list.

The middle name is $\left[\dfrac{7 + 10}{2}\right] = [8.5] = 9$ Saludo.

Robinson is before Saludo, so the list reduces to

7 Leung

8 Robinson.

Robinson is before Saludo so it cannot be in the second part of the list. We therefore delete the pivot 9 and everything after it.

The middle name is $\left[\dfrac{7 + 8}{2}\right] = [7.5] = 8$ Robinson.

The search is complete.
Robinson has been found at 8.

b The middle name is $\left[\dfrac{10 + 1}{2}\right] = [5.5] = 6$ Laing.

Davies is before Laing, so the list reduces to

1 Bennett

2 Blackstock

3 Brown

4 Ebenezer

5 Fowler.

Delete the pivot, 6 and the second half of the list.

The middle name is $\left[\dfrac{1 + 5}{2}\right] = [3] = 3$ Brown.

Davies is after Brown, so the lists reduces to

4 Ebenezer

5 Fowler.

Delete the pivot, 3, and everything before it.

The middle name is $\left[\dfrac{4 + 5}{2}\right] = [4.5] = 5$ Fowler.

Davies is before Fowler, so the list reduces to

4 Ebenezer.

Delete the pivot, 5, and everything after it.

The list is now only one item and this item is not Davies.
We conclude that Davies is not in the list.

Example 13

Use the binary search algorithm to locate the number 12 in the list opposite.

1 2
2 3
3 5
4 7
5 11
6 13
7 17
8 19
9 23
10 29
11 31

The middle number is in position number $\left[\dfrac{1+11}{2}\right] = 6$

The 6th number is 13.

12 comes before 13, so the list reduces to

1 2

2 3

3 5

4 7

5 11

> The algorithm finds the *position* of each pivot, then compares the pivot with the intended target.

The middle number is in position number $\left[\dfrac{1+5}{2}\right] = 3$

The 3rd number is 5.

12 comes after 5, so the list reduces to

4 7

5 11

The middle number is in position number $\left[\dfrac{4+5}{2}\right] = [4.5] = 5$

The 5th number is 11.

12 comes after 11, so the list reduces to nothing.

We conclude that 12 is not in the list.

> We need to delete the pivot, the 5th number, and everything before it. This deletes everything in the current list. We have nothing left.

Exercise 1D

1 Use the binary search algorithm to try to locate

   a Connock,   **b** Walkey,   **c** Peabody.

   in the list below.

   1 Berry

   2 Connock

   3 Ladley

   4 Sully

   5 Tapner

   6 Walkey

   7 Ward

   8 Wilson

2 Use the binary search algorithm to try to locate

    a 21,   **b** 5

in the list below.

| **1** 3 | **3** 7 | **5** 10 | **7** 15 | **9** 18 | **11** 21 |
| **2** 4 | **4** 9 | **6** 13 | **8** 17 | **10** 20 | **12** 24 |

3 Use the binary search algorithm to try to locate

    a Fredco,   **b** Matt,   **c** Elliot

in the list below.

1 Adam	**6** Emily	**11** Katie	**16** Miranda
2 Alex	**7** Fredco	**12** Leo	**17** Oli
3 Des	**8** George	**13** Lottie	**18** Ramin
4 Doug	**9** Harry	**14** Louis	**19** Saul
5 Ed	**10** Jess	**15** Matt	**20** Simon

4 The 26 letters of the English alphabet are listed, in order.

    a Apply the binary search algorithm to locate the letter P.

    b What is the maximum number of iterations needed to locate any letter?

5 The binary search algorithm is applied to an ordered list of n items.
Determine the maximum number of iterations needed
when n is equal to

    a 100   **b** 1000   **c** 10 000.

> You may find it helpful to record the maximum length of the list after each iteration.

1.6 You need to be able to implement the three bin packing algorithms and be aware of their limitations.

■ Bin packing refers to a whole class of problems.

■ The easiest is to imagine stacking boxes of fixed width a and length b, but varying height, into bins of width a and length b, using the minimum number of bins.

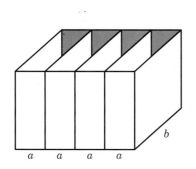 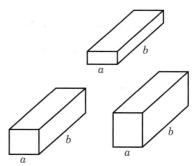

■ Similar problems could be: loading cars onto a ferry with several lanes of equal length, a plumber needing to cut sections from lengths of copper pipe, or placing music tracks onto a set of CDs.

The lower bound

Example 14

Nine boxes of fixed cross section have heights, in metres, as follows.

0.3, 0.7, 0.8, 0.8, 1.0, 1.1, 1.1, 1.2, 1.5

They are to be packed into bins with the same fixed cross section and height 2 m.
Determine the lower bound for the number of bins needed.

$$0.3 + 0.7 + 0.8 + 0.8 + 1.0 + 1.1$$
$$+ 1.1 + 1.2 + 1.5 = 8.5 \, m$$

$$\frac{8.5}{2} = 4.25 \text{ bins}$$

So a minimum of 5 bins will be needed.

Sum the heights and divide by the bin size. You must always round *up* to determine the lower bound.

Note It may not be possible to pack these boxes into 5 bins. That will depend on the numbers. All that the lower bound is telling us, is that *at least* five bins will be needed.

■ At present there is no known algorithm that will always give an optimal solution.

■ With small amounts of data it is often possible to 'spot' an optimal answer.

■ There are three bin packing algorithms in common use: **first-fit**, **first-fit decreasing** and **full-bin packing**.

First-fit algorithm

1 Take the items **in the order given**.

2 Place each item in the first available bin that can take it. Start from bin 1 each time.
Advantage: It is quick to do.
Disadvantage: It is not likely to lead to a good solution.

Example 15

Use the first-fit algorithm to pack the following items into bins of size 20. (The numbers in brackets are the size of the item.) State the number of bins used and the amount of wasted space.

A(8)   B(7)   C(14)   D(9)   E(6)   F(9)   G(5)   H(15)   I(6)   J(7)   K(8)

Bin 1:	A(8)	B(7)	G(5)
Bin 2:	C(14)	E(6)	
Bin 3:	D(9)	F(9)	
Bin 4:	H(15)		
Bin 5:	I(6)	J(7)	
Bin 6:	K(8)		

This used 6 bins and there are
$2 + 5 + 7 + 12 = 26$ units of waste of space.

A(8) goes into bin 1,   leaving space of 12.
B(7) goes into bin 1,   leaving space of 5.
C(14) goes into bin 2, leaving space of 6.
D(9) goes into bin 3,   leaving space of 11.
E(6) goes into bin 2,   leaving space of 0.
F(9) goes into bin 3,   leaving space of 2.
G(5) goes into bin 1,   leaving space of 0.
H(15) goes into bin 4, leaving space of 5.
I(6) goes into bin 5,   leaving space of 14.
J(7) goes into bin 5,   leaving space of 7.
K(8) goes into bin 6,   leaving space of 12.

First-fit decreasing algorithm

1 Reorder the items so that they are in descending order.

2 Apply the first-fit algorithm to the reordered list.

Advantages: You usually get a fairly good solution.
It is easy to do.
Disadvantage: You may not get an optimal solution.

> In the examination you may be asked to apply a sorting algorithm to do this.

> The algorithm is sometimes referred to as the mathematician's way of packing a suitcase – put the big things in first and the socks last!

Example 16

Apply the first-fit decreasing algorithm to the data given in Example 15.

Sort the data into descending order.
H(15)   C(14)   D(9)   F(9)   A(8)   K(8)
B(7)   J(7)   E(6)   I(6)   G(5)

Bin 1:	H(15)	G(5)	
Bin 2:	C(14)	E(6)	
Bin 3:	D(9)	F(9)	
Bin 4:	A(8)	K(8)	
Bin 5:	B(7)	J(7)	I(6)

This used 5 bins and there are
2 + 4 = 6 units of wasted space.

H(15) goes into bin 1, leaving space of 5.
C(14) goes into bin 2, leaving space of 6.
D(9) goes into bin 3,   leaving space of 11.
F(9) goes into bin 3,   leaving space of 2.
A(8) goes into bin 4,   leaving space of 12.
K(8) goes into bin 4,   leaving space of 4.
B(7) goes into bin 5,   leaving space of 13.
J(7) goes into bin 5,   leaving space of 6.
E(6) goes into bin 2,   leaving space of 0.
I(6) goes into bin 5,   leaving space of 0.
G(5) goes into bin 1,   leaving space of 0.

Full-bin packing

1 Use observation to find combinations of items that will fill a bin. Pack these items first.

2 Any remaining items are packed using the first-fit algorithm.

Advantage: You usually get a good solution.
Disadvantage: It is difficult to do, especially when the numbers are plentiful and awkward.

Example 17

A(8)   B(7)   C(10)   D(11)   E(13)   F(17)   G(4)   H(6)   I(12)   J(14)   K(9)

The items above are to be packed in bins of size 25.

a Determine the lower band for the number of bins.

b Apply the full-bin algorithm.

c Is your solution optimal? Give a reason for your answer.

a   Lower band = 111 ÷ 25 = 4.44,
so 5 bins are needed.

b 17 + 8 = 25

13 + 12 = 25

14 + 11 = 25

so a solution is

Bin 1:   F(17)   A(8)

Bin 2:   E(13)   I(12)

Bin 3:   J(14)   D(11)

Bin 4:   B(7)   C(10)   G(4)

Bin 5:   H(6)   K(9)

> The first three bins are full bins.
> We now apply the first-fit algorithm to the remainder.
> B(7) goes into bin 4,   leaving space of 18.
> C(10) goes into bin 4, leaving space of 8.
> G(4) goes into bin 4,   leaving space of 4.
> H(6) goes into bin 5,   leaving space of 19.
> K(9) goes into bin 5,   leaving space of 10.

c   The lower bound is 5 and 5 bins were used, so the solution is optimal.

Example 18

A plumber needs to cut the following lengths of copper pipe. (Lengths are in metres.)

A(0.8)   B(0.8)   C(1.4)   D(1.1)   E(1.3)   F(0.9)   G(0.8)   H(0.9)   I(0.8)   J(0.9)

The pipe comes in lengths of 2.5 m.

a Calculate the lower bound of the number of 2.5 m lengths needed.

b Use the first-fit decreasing algorithm to determine how the required lengths may be cut from the 2.5 m lengths.

c Use full-bin packing to find an optimal solution.

a $\dfrac{0.8 + 0.8 + 1.4 + 1.1 + 1.3 + 0.9 + 0.8 + 0.9 + 0.8 + 0.9}{2.5} = 3.88$

So at least 4 lengths are required.

> Since a sort was not asked for, this can be done by inspection.

b   Sorting into descending order

C(1.4), E(1.3), D(1.1), F(0.9), H(0.9), J(0.9), A(0.8), B(0.8), G(0.8), I(0.8)

Bin 1:   C(1.4)   D(1.1)

Bin 2:   E(1.3)   F(0.9)

Bin 3:   H(0.9)   J(0.9)

Bin 4:   A(0.8)   B(0.8)   G(0.8)

Bin 5:   I(0.8)

> C goes into bin 1, leaving space of 1.1.
> E goes into bin 2, leaving space of 1.2.
> D goes into bin 1, leaving space of 0.
> F goes into bin 2, leaving space of 0.3.
> H goes into bin 3, leaving space of 1.6.
> J goes into bin 3, leaving space of 0.7.
> A goes into bin 4, leaving space of 1.7.
> B goes into bin 4, leaving space of 0.9.
> G goes into bin 4, leaving space of 0.1.
> I goes into bin 5,   leaving space of 1.7.

c   By inspection

C(1.4) + D(1.1) = 2.5

F(0.9) + A(0.8) + B(0.8) = 2.5

J(0.9) + G(0.8) + I(0.8) = 2.5

A full-bin solution is

Bin 1:   C(1.4)   D(1.1)

Bin 2:   F(0.9)   A(0.8)   B(0.8)

Bin 3:   J(0.9)   G(0.8)   I(0.8)

Bin 4:   E(1.3)   H(0.9)

> In part **a** we found that at least 4 bins would be needed, so this solution is optimal since it uses 4 bins.

Exercise 1E

1
   18   4   23   8   27   19   3   26   30   35   32

The above items are to be packed in bins of size 50.

a Calculate the lower bound for the number of bins.

b Pack the items into the bins using
   i the first-fit algorithm,   **ii** the first-fit decreasing algorithm,   **iii** the full-bin algorithm.

2 Laura wishes to record the following television programmes onto DVDs, each of which can hold up to 3 hours of programmes.

Programme	A	B	C	D	E	F	G	H	I	J	K	L	M
Length (minutes)	30	30	30	45	45	60	60	60	60	75	90	120	120

a Apply the first-fit algorithm, in the order A to M, to determine the number of DVDs that need to be used. State which programmes should be recorded on each disc.

b Repeat part **a** using the first-fit decreasing algorithm.

c Is your answer to part **b** optimal? Give a reason for your answer.

Laura finds that her DVDs will only hold up to 2 hours of programmes.

d Use the full-bin algorithm to determine the number of DVDs she needs to use. State which programmes should be recorded on each disc.

3 A small ferry has three car lanes, each 30 m long. There are 10 vehicles waiting to use the ferry.

	Vehicle	Length (m)
A	car	4 m
B	car + trailer	7 m
C	lorry	13 m
D	van	6 m
E	lorry	13 m

	Vehicle	Length (m)
F	car	4 m
G	lorry	12 m
H	lorry	14 m
I	van	6 m
J	lorry	11 m

a Apply the first-fit algorithm, in the order A to J. Is it possible to load all the vehicles using this method?

b Apply the first-fit decreasing algorithm. Is it possible to load all the vehicles using this method?

c Use full-bin packing to load all of the vehicles.

4 The ground floor of an office block is to be fully recarpeted, with specially made carpet which incorporate the firm's logo. The carpet comes in rolls of 15 m.

The following lengths are required.

A 3 m   D 4 m   G 5 m   J 7 m
B 3 m   E 4 m   H 5 m   K 8 m
C 4 m   F 4 m   I 5 m   L 8 m

Determine how the lengths should be cut from the rolls using

a the first-fit algorithm A to L,

b the first-fit decreasing algorithm,

c full-bin packing.

In each case, state the number of rolls used and the amount of wasted carpet.

5 Eight computer programs need to be copied onto 40 MB discs.

The size of each program is given below.

Program	A	B	C	D	E	F	G	H
Size (MB)	8	16	17	21	22	24	25	25

a Use the first-fit decreasing algorithm to determine which programs should be recorded onto each disc.

b Calculate a lower bound for the number of discs needed.

c Explain why it is not possible to record these programs on the number of discs found in part **b**.

Hint: Consider the programs over 20 MB in size.

Mixed exercise 1F

Photocopy masters are available for the questions marked * in this exercise.

1 Use the bubble-sort algorithm to sort, in ascending order, the list:

27   15   2   38   16   1

giving the state of the list at each stage.   **E**

2 **a** Use the bubble-sort algorithm to sort, in descending order, the list:

25   42   31   22   26   41

giving the state of the list on each occasion when two values are interchanged.

b Find the *maximum* number of interchanges needed to sort a list of six pieces of data using the bubble-sort algorithm.   **E**

3   8   4   13   2   17   9   15

This list of numbers is to be sorted into ascending order.

Perform a quick sort to obtain the sorted list, giving the state of the list after each rearrangement.   **E**

4   111   103   77   81   98   68   82   115   93

a The list of numbers above is to be sorted into descending order. Perform a quick-sort to obtain the sorted list, giving the state of the list after each rearrangement and indicating the pivot elements used.

b i Use the first-fit decreasing bin-packing algorithm to fit the data into bins of size 200.

ii Explain how you decided in which bin to place the number 77.   **E**

5 Trishna wishes to video eight television programmes. The lengths of the programmes, in minutes, are:

75   100   52   92   30   84   42   60

Trishna decides to use 2-hour (120 minute) video tapes only to record all of these programmes.

a Explain how to use a first-fit decreasing bin-packing algorithm to find the solution that uses the fewest tapes and determine the total amount of unused tape.

b Determine whether it is possible for Trishna to record an additional two 25-minute programmes on these 2-hour tapes, without using another video tape.   **E**

LONGLEY PARK SIXTH FORM COLLEGE
HORNINGLOW ROAD
SHEFFIELD
S5 5SG

6 A DIY enthusiast requires the following 14 pieces of wood as shown in the table.

Length in metres	0.4	0.6	1	1.2	1.4	1.6
Number of pieces	3	4	3	2	1	1

The DIY store sells wood in 2 m and 2.4 m lengths. He considers buying six 2 m lengths of wood.

a Explain why he will not be able to cut all of the lengths he requires from these six 2 m lengths.

b He eventually decides to buy 2.4 m lengths. Use a first-fit decreasing bin-packing algorithm to show how he could use six 2.4 m lengths to obtain the pieces he requires.

c Obtain a solution that only requires five 2.4 m lengths.　　　　　　　**E**

7 *Note:* This question uses the modulus function. If $x \neq y$, $|x - y|$ is the positive difference between x and y, e.g. $|5 - 6.1| = 1.1$. The algorithm described by the flow chart below is to be applied to the five pieces of data below.

$U(1) = 6.1, U(2) = 6.9, U(3) = 5.7, U(4) = 4.8, U(5) = 5.3$

a Obtain the final output of the algorithm using the five values given for $U(1)$ to $U(5)$.

b In general, for any set of values $U(1)$ to $U(5)$, explain what the algorithm achieves.

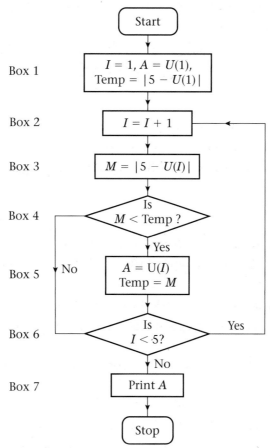

c If Box 4 in the flow chart is altered to

'Is $M > $ Temp?'

state what the algorithm now achieves.　　　　　　　**E**

Summary of key points

1 Algorithms can be given in words or flowcharts.

2 Unordered lists can be sorted using a bubble sort or a quick sort.

3 In a **bubble sort**, you compare adjacent items in a list.
  - If they are in order, leave them.
  - If they are not in order, swap them.
  - The list is in order when a pass is completed without any swaps.

4 In a **quick sort**, you select a pivot and then split the items into two sub-lists: those less than the pivot and those greater than the pivot.
  - Keep doing this to each resulting sub-list.

5 In a quick sort, the item at the mid-point of the list is chosen as the pivot.
  - Items less than the pivot are written down, keeping their order.
  - The pivot is written next.
  - Items greater than the pivot are written down, keeping their order.

6 A **binary search** will search an **ordered list** to find out whether an item is in the list. If it is in the list, it will locate its position in the list.

7 In a binary search, the pivot is the middle item in the list. If the target item is not the pivot, the pivot and half the list are discarded. The list length halves at each pass.

8 The middle of n items is found by $\left[\dfrac{1 + n}{2}\right]$, rounding up if necessary.

9 The three bin packing algorithms are: first-fit, first-fit decreasing, and full-bin.

10 - First-fit algorithm takes items in the order given.
   - First-fit decreasing algorithm requires the items to be in descending order before applying the algorithm.
   - Full-bin packing uses inspection to select items that will combine to fill bins. Remaining items are packed using the first-fit algorithm.

11 The three bin packing algorithms have advantages and disadvantages.

	Advantage	Disadvantage
First-fit	Quick to do	Not likely to lead to a good solution
First-fit decreasing	Usually a good solution Easy to do	May not get an optimal solution
Full-bin	Usually a good solution	Difficult to do, especially when lots of numbers or awkward numbers

After completing this chapter you should:

1 know how graphs and networks can be used to create mathematical models

2 know some basic terminology used in graph theory

3 know some special types of graph

4 understand how graphs and networks can be represented using matrices.

Graphs and networks

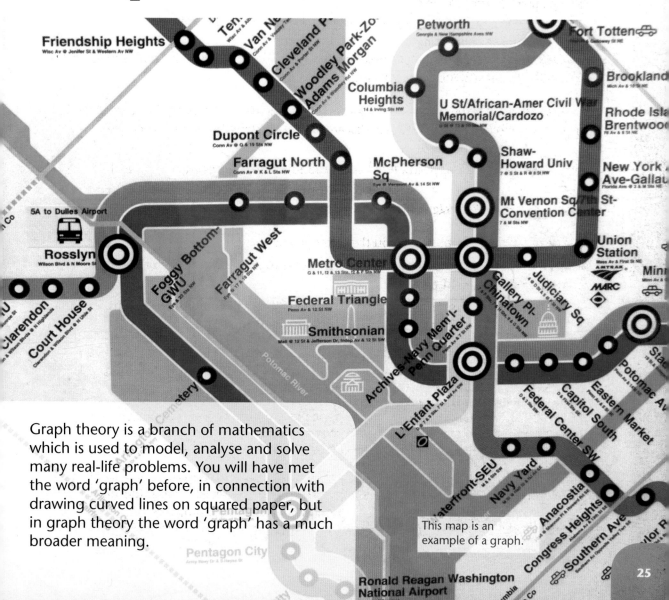

Graph theory is a branch of mathematics which is used to model, analyse and solve many real-life problems. You will have met the word 'graph' before, in connection with drawing curved lines on squared paper, but in graph theory the word 'graph' has a much broader meaning.

This map is an example of a graph.

2.1 **You need to know how graphs and networks can be used to create mathematical models.**

■ A **graph** consists of points (called **vertices** or **nodes**) which are connected by lines (**edges** or **arcs**).

■ If a graph has a number associated with each edge (usually called its weight), then the graph is known as a **weighted graph** or **network**.

Example 1

a Why is this a graph?

b What are **i** the vertices, **ii** the edges?

c What is the purpose of this graph?

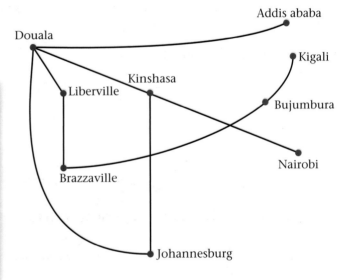

a	It has vertices connected by edges.
b **i**	Cities seved by an airline
ii	Flight routes connecting cities
c	To make route planning on the airline possible.
	It does not show any distances or the correct 'geographical' position of the cities.

Example 2

a Why is this a graph?

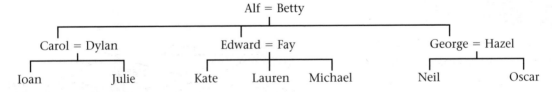

b What does it represent?

c What do **i** the vertices, **ii** the edges represent?

a	It has vertices connected by edges.
b	A family tree.
c	**i** The people in the family.
	ii The relationships between the people.

Example 3

The editor of a newspaper has to assign five reporters, Amro, Bavika, Chi, David and Erin, to cover five events: hospital opening, movie premiere, celebrity wedding, royal visit and 110th birthday party.

Amro could be assigned to the movie premiere or the celebrity wedding.

Bavika could be assigned to the hospital opening, movie premiere or 110th birthday.

Chi could be assigned to the royal visit, celebrity wedding or 110th birthday.

David could be assigned to the hospital opening, movie premiere or royal visit.

Erin could be assigned to the celebrity wedding, royal visit or 110th birthday.

Complete this graph to model this information.

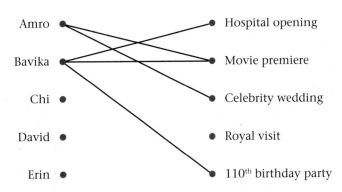

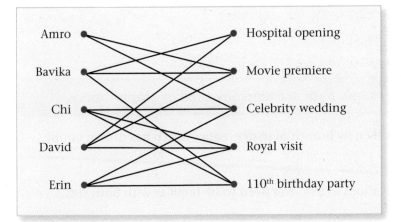

Here the vertices represent the reporters and the events, and the edges indicate the possible assignments.

Example 4

The manager of a camp site needs to provide a water supply to standpipes located at various points on the site.

a What do the vertices represent?

b What do the edges represent?

c What do the numbers represent?

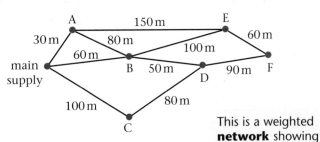

This is a weighted **network** showing a simplified view of the situation.

a	the standpipes
b	the possible supply routes
c	the lengths of each supply route

Example 5

This network represents a project with six activities, A, B, C, D, E and F.
Interpret the network.

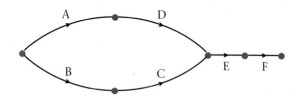

Activities A and B can start straightaway.
Activity C cannot start until B is completed.
Activity D cannot start until A is completed.
Activity E cannot start until both D and C are completed.
Activity F cannot start until E is completed.

2.2 You need to be familiar with some basic terminology used in graph theory.

■ Decision mathematics is a relatively new branch of mathematics which has grown out of several fields of study.

■ There are often two different notations in use. You need to be familiar with both sets.

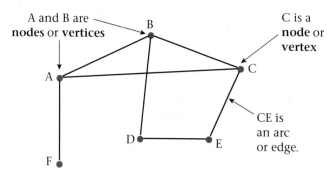

In the **graph G**, above

■ the **vertices** (or **nodes**) are: A, B, C, D, E and F
(this list of vertices is sometimes called the **vertex set**),

■ the **edges** (or **arcs**) are: AB, AC, AF, BC, BD, CE and DE
(this list of edges is sometimes called the **edge set**).
Note: the intersection of AC and BD is not a vertex.

■ A **subgraph** of G is a graph, each of whose vertices belongs to G and each of whose edges belongs to G. It is simply a part of the original graph.

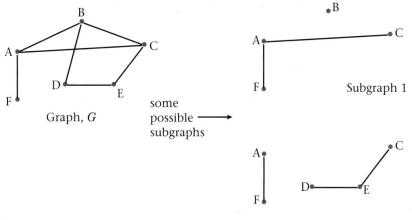

Graph, G

some possible ⟶ subgraphs

Subgraph 1

Subgraph 2

■ The **degree** or **valency** or **order** of a vertex is the number of edges incident to it.

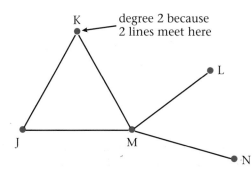

degree 2 because 2 lines meet here

Vertex	Degree
J	2
K	2
L	1
M	4
N	1

All three words mean the same thing.

In any graph the sum of the degrees will be precisely equal to 2 × the number of edges. This is because, in finding the sum of degrees, we are counting each end of each edge.

This is known as the *Handshaking Lemma*. (See Section 4.1.)

■ If the degree of a vertex is even, we say it has **even degree**, so J, K and M have even degree.
Similarly vertices L and N have odd degree.

■ A **path** is a finite sequence of edges, such that the end vertex of one edge in the sequence is the start vertex of the next, and in which no vertex appears more than once.

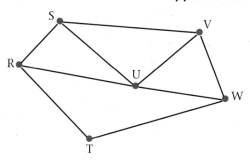

For example, in the graph above, one possible path is R S U W T and another is U V W T R S.

■ A **walk** is a path in which you are permitted to return to vertices more than once. For example in the graph above, a walk could be R U V S R T W U V.

■ A **cycle** (or **circuit**) is a closed 'path', i.e. the end vertex of the last edge is the start vertex of the first edge.
For example in the graph above, a possible cycle is R U V W T R.

■ Two vertices are **connected** if there is a path between them. A **graph** is **connected** if all its vertices are connected.

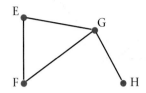

This is a **connected** graph.
A path can be found between any two vertices.

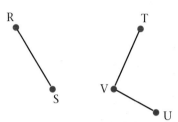

This shows a graph that is **not connected**.
There is no path from R to V, for example.

■ A **loop** is an edge that starts and finishes at the same vertex.

This contains a **loop** from C to C.

loop

■ A **simple graph** is one in which there are no loops and not have more than one edge connecting any pair of vertices.

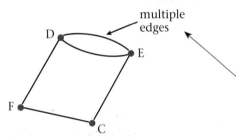

multiple edges

The graph above is not a simple graph because it contains a loop.

This is not a simple graph because it has two edges from D to E.

■ If the edges of a graph have a direction associated with them they are known as **directed edges** and the graph is known as a **digraph**.

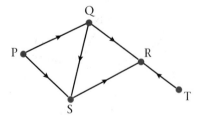

Exercise 2A

1 Draw a connected graph with

   a one vertex of degree 4 and 4 vertices of degree 1,

   b three vertices of degree 2, one of degree 3 and one of degree 1,

   c two vertices of degree 2, two of degree 3 and one of degree 4.

2 Which of the graphs below are not simple?

a

b
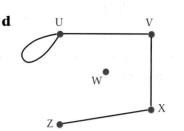

c
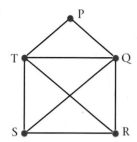

d

3 In question 2, which graphs are not connected?

4 From the graph, state

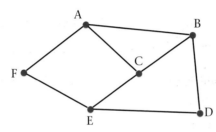

> There are many correct answers to these questions.

a four paths from F to D,

b a cycle passing through F and D,

c the degree of each vertex.

Use the graph to

d draw a subgraph,

e confirm the handshaking lemma (that the sum of the degrees is equal to twice the number of edges).

> A lemma is a mathematical fact used as a stepping stone to more important results.

5 a Repeat question 4 parts **c**, **d** and **e** using this graph.

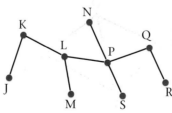

b Confirm that there is only one path between any two vertices.

6 Show that it is possible to draw a graph with

a an even number of vertices of even degree,

b an odd number of vertices of even degree.

It is not possible to draw a graph with an odd number of vertices of odd degree. Explain why not.

> Use the handshaking lemma.

7 Five volunteers, Ann, Brian, Conor, Dave and Eun Jung are going to run a help desk from Monday to Friday next week.

One person is required each day.

Ann is available on Tuesday and Friday,
Brian is available on Wednesday,
Conor is available on Monday, Thursday and Friday,
Dave is available on Monday, Wednesday and Thursday,
Eun Jung is available on Tuesday, Wednesday and Thursday.

Draw a graph to model this situation.

Look at Example 3.

8 A project consists of six activities A, B, C, D, E and F.

A and B can start immediately, but C cannot start until A is completed.
D cannot start until both B and C are complete, E cannot start until D is complete and F cannot start until E is complete.

Draw a digraph to model this situation.

Look at Example 5.

2.3 You need to know some special types of graph.

■ A **tree** is a connected graph with no cycles.

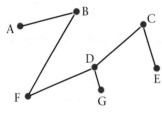

A tree

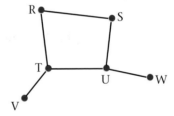

**This is not a tree.
It contains a cycle R T U S R.**

■ A **spanning tree** of a graph, G, is a subgraph which includes all the vertices of G and is also a tree.

For example, starting with this graph, G:

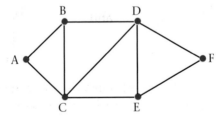

Possible spanning trees are

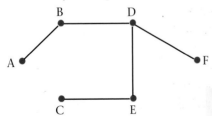

or

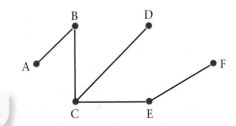

There are *many* other spanning trees

■ A **bipartite** graph consists of two sets of vertices, X and Y. The edges only join vertices in X to vertices in Y, not vertices within a set.

You saw an example of a bipartite graph in Example 3. Another one is drawn below.

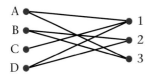

Notice that each edge connects a left hand vertex set to a right hand vertex set. There are no edges that connect vertices in the same set.

■ A **complete graph** is a graph in which every vertex is directly connected by an edge to each of the other vertices. If the graph has n vertices the connected graph is denoted by k_n.

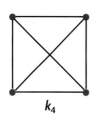

k_4

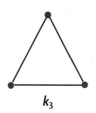

k_3

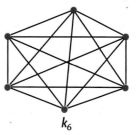

k_6

■ A **complete bipartite graph** (denoted by $k_{r,\,s}$) in which there are r vertices in set X and s vertices in set Y.

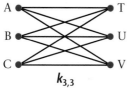

$k_{3,3}$

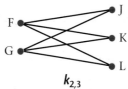

$k_{2,3}$

■ **Isomorphic graphs** are graphs that show the same information but are drawn differently.

For example

is isomorphic to

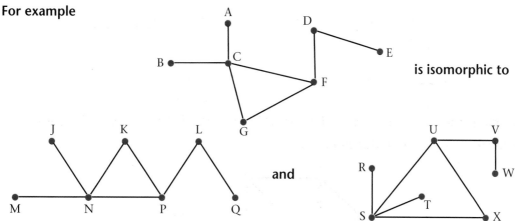

and

For two graphs to be isomorphic, they must have the same number of vertices of the same degree, but these vertices must also be connected together in the same way.

If graphs are isomorphic it is possible to pair equivalent vertices in this case.
A can be paired with J in the first graph and R in the second.
B can be paired with M in the first graph and T in the second.
C can be paired with N in the first graph and S in the second.
D can be paired with L in the first graph and V in the second.
E can be paired with Q in the first graph and W in the second.
F can be paired with P in the first graph and U in the second.
G can be paired with K in the first graph and X in the second.

2.4 **You should understand how graphs and networks can be represented using matrices.**

■ An **adjacency matrix** records the number of direct links between vertices.

■ A **distance matrix** records the weights on the edges. Where there is no edge, we write '−'.

Example 6

Use an adjacency matrix to represent this graph.

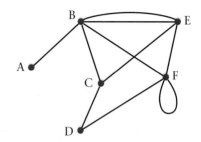

	A	B	C	D	E	F
A	0	1	0	0	0	0
B	1	0	1	0	2	1
C	0	1	0	1	1	0
D	0	0	1	0	0	1
E	0	2	1	0	0	1
F	0	1	0	1	1	2

You should be able to write down the adjacency matrix given the graph, and draw a graph given the adjacency matrix.

This indicates that there are 2 direct connections between B and E.

This indicates a loop from F to F. It could be travelled in either direction, and hence counts as 2.

Example 7

Use a distance matrix to represent this network.

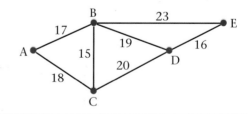

	A	B	C	D	E
A	−	17	18	−	−
B	17	−	15	19	23
C	18	15	−	20	−
D	−	19	20	−	16
E	−	23	−	16	−

You should be able to write down the distance matrix given the network and draw the network given the distance matrix.

You will not have to construct matrices for non-directed networks with loops or multiple edges.

Notice that the matrix is symmetrical about the leading diagonal (top left to bottom right).

Example 8

Use a distance matrix to represent this directed network.

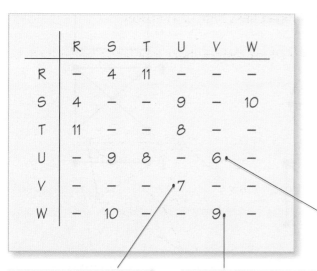

	R	S	T	U	V	W
R	–	4	11	–	–	–
S	4	–	–	9	–	10
T	11	–	–	8	–	–
U	–	9	8	–	6	–
V	–	–	–	7	–	–
W	–	10	–	–	9	–

The matrix will not be symmetrical about the leading diagonal for a directed network.

This indicates a direct link of weight 6 from U to V.

This indicates a direct link of 7 from V to U.

This shows a direct link of 9 from W to V.

Exercise 2B

1 State which of the following graphs are trees.

a

b

c

d

2

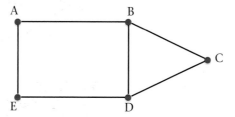

There are 11 spanning trees for the graph above. See how many you can find.

3 Draw k_8.

What is the degree of each vertex in the graph k_n?

4 Draw $k_{3,4}$

How many edges would be in $k_{n,m}$?

5 Which of these graphs are isomorphic?

a

b

c

d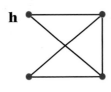

e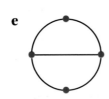

f

g

h

i

6

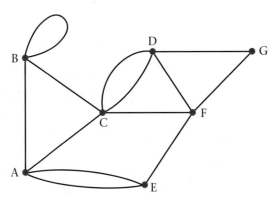

Use an adjacency matrix to represent the graph above.

7 Draw a graph corresponding to each adjacency matrix.

a

	A	B	C	D	E
A	0	1	0	1	0
B	1	0	1	1	1
C	0	1	0	2	0
D	1	1	2	0	1
E	0	1	0	1	0

b

	A	B	C	D
A	0	1	0	1
B	1	0	0	1
C	0	0	2	1
D	1	1	1	0

c

	A	B	C	D
A	0	1	0	1
B	1	0	1	1
C	0	1	0	1
D	1	1	1	0

d

	A	B	C	D
A	0	2	0	1
B	2	0	1	0
C	0	1	0	1
D	1	0	1	0

8 Which graphs in Question **5** could be described by the adjacency matrices in Question **7c** and **d**?

9 Draw the network corresponding to each distance matrix.

a

	A	B	C	D	E
A	–	21	–	20	23
B	21	–	17	23	–
C	–	17	–	18	41
D	20	23	18	–	22
E	23	–	41	22	–

b

	A	B	C	D	E	F
A	–	–	–	–	15	8
B	–	–	9	13	17	11
C	–	9	–	8	–	–
D	–	13	8	–	10	–
E	15	17	–	10	–	–
F	8	11	–	–	–	–

10

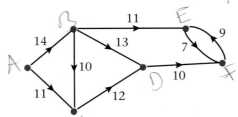

Use a distance matrix to represent the directed network above.

Mixed exercise 2C

1 A group of 6 children, Ahmed, Bronwen, Di, Eddie, Fiona and Gary, were asked which of 6 sports, football, swimming, cricket, tennis, hockey and rugby, they enjoyed playing. The results are shown in the table below.

Ahmed	football	cricket	rugby		
Bronwen	swimming	cricket	hockey		
Di	football	tennis	hockey		
Eddie	football	cricket	tennis	hockey	rugby
Fiona	tennis	swimming	hockey		
Gary	football	swimming	hockey		

Draw a bipartite graph to show this information.

2 A project involves eight activities A, B, C, D, E, F, G and H, some of which cannot be started until others are completed. The table shows the tasks that need to be completed before the activity can start. For example, activity F cannot start until both B and E are completed.

Activity	Activity that must be completed
A	–
B	–
C	–
D	A
E	A
F	B E
G	D F
H	C G

Draw a digraph to represent this information.

3 **a** Draw a graph with eight vertices, all of degree 1.

b Draw a graph with eight vertices, all of degree 2, so that the graph is

    i   connected and simple

    ii   not connected and simple

    iii not connected and not simple.

4 Use a distance matrix to represent the network below.

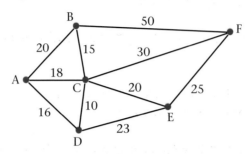

5 Find two spanning trees for the graph in Question **4**.

6 Write down a formula connecting the number of edges, E, in a spanning tree with V vertices.

Summary of key points

1 A **graph** consists of **vertices (nodes)** which are connected by **edges (arcs)**.

2 A **subgraph** is part of a graph.

3 If a graph has a number associated with each edge (its **weight**) then the graph is a **weighted graph (network)**.

4 The **degree** or **valency** (or **order**) of a vertex is the number of edges incident to it.

5 A **path** is a finite sequence of edges, such that the end vertex of one edge in the sequence is the start vertex of the next, and in which no vertex appears more than once.

6 A **walk** is a path in which you are permitted to return to vertices more than once.

7 A **cycle** (**circuit**) is a closed 'path', i.e. the end vertex of the last edge is the start vertex of the first edge.

8 Two vertices are **connected** if there is a path between them. A graph is connected if all its vertices are connected.

9 A **loop** is an edge that starts and finishes at the same vertex.

10 A **simple graph** is one in which there are no loops and not more than one edge connecting any pair of vertices.

11 If the edges of a graph have a direction associated with them they are known as **directed edges** and the graph is known as a **digraph**.

12 A **tree** is a connected graph with no cycles.

13 A **spanning tree** of a graph, G, is a subgraph which includes all the vertices of G and is also a tree.

14 A **bipartite graph** consists of two sets of vertices, X and Y. The edges only join vertices in X to vertices in Y, not vertices within a set.

15 A **complete graph** is a graph in which every vertex is directly connected by an edge to each of the other vertices. If the graph has n vertices the connected graph is denoted k_n.

16 A **complete bipartite graph** (denoted $k_{r, s}$) is a graph in which there are r vertices in set X and s vertices in set Y.

17 **Isomorphic graphs** show the same information but are drawn differently.

18 An **adjacency matrix** records the number of direct links between vertices.

19 A **distance matrix** records the weights on the edges. Where there is no weight, this is indicated by '$-$'.

3

After completing this chapter you should be able to:

1 use Kruskal's algorithm to find a minimum spanning tree

2 use Prim's algorithm on a network to find a minimum spanning tree

3 apply Prim's algorithm to a distance matrix

4 use Dijkstra's algorithm to find the shortest path between two vertices in a network.

Algorithms on networks

Each telephone needs to be connected into the system. As long as all the telephones are in the system, everyone can talk to everyone else. There is no need for a direct line between each of the telephones.

The algorithms in this chapter can be used to find the shortest distance between two nodes, or the shortest way of linking all the nodes.

3.1 You can use Kruskal's algorithm to find a minimum spanning tree.

■ A **minimum spanning tree (MST)** is a spanning tree such that the total length of its arcs (edges) is as small as possible. (An MST is sometimes called a **minimum connector**.)

■ Kruskal's algorithm finds the shortest, cheapest or fastest way of linking all the nodes into one system.

■ Here is Kruskal's algorithm.

> You may need to check back in Chapter 2 to make sure that you understand the meaning of the words **arc**, **weight**, **tree**, **cycle**, **vertices**.

1 Sort all the **arcs** (edges) into *ascending order of* **weight**.

2 Select the arc of *least weight* to *start the* **tree**.

3 Consider the next arc of least weight.
  • If it would form a **cycle** with the arcs already selected, reject it.
  • If it does not form a cycle, add it to the tree.
  If there is a choice of equal arcs, consider each in turn.

4 Repeat step 3 until all **vertices** are connected.

Example 1

Use Kruskal's algorithm to find a minimum spanning tree (minimum connector) for this network. List the arcs in the order that you consider them. State the weight of your tree.

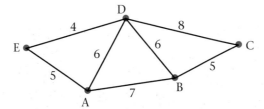

> In a network of *n* vertices a spanning tree will always have $(n-1)$ arcs. In this case there will be 4 arcs in the spanning tree.

By inspection the order of the arcs is

DE(4), AE(5), BC(5), AD(6), BD(6), AB(7), CD(8)

Start with DE.

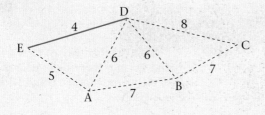

> AE and BC could have been written in either order as they are both 5. In the same way, AD and BD could have been written in either order.

Add AE.

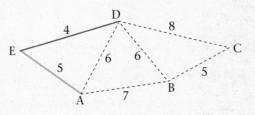

> You do not need to draw each of these diagrams. The list of arcs, in order, with your decision about rejecting or adding them, is sufficient to make your method clear.

Add BC.

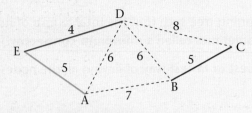

Reject AD (it would make the cycle AEDA).
Add BD.

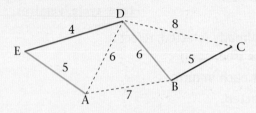

This is the minimum spanning tree. It includes all the vertices.

Its weight is 5 + 4 + 6 + 5 = 20

All the vertices are now connected so we can stop.

If you continued to work through the list of arcs, each would be rejected because they would make cycles.

Example 2

Use Kruskal's algorithm and show that there are four minimum spanning trees for this network. State their weight.

You may find it helpful to draw out the tree as you go. It makes it easier to check for cycles.

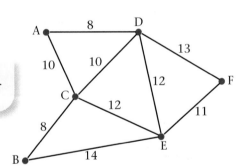

Order of arcs:

length 8:   AD and BC
length 10: AC and CD
length 11: EF
length 12: CE and DE
length 13: DF
length 14: BE

Spanning tree

Start with AD.

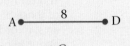

Remember that we can choose AD and BC in either order, AC and CD in either order and CE and DE in either order.

Add BC.

Add AC.

Reject CD (it forms a cycle ACDA).

Add EF.

Add CE.

One solution is

(Using AC and CE.)

The other 3 solutions are

(Using AC and DE.)   (Using CD and CE.)

(Using CD and DE.)

The weight of each tree is 8 + 8 + 10 + 11 + 12 = 49.

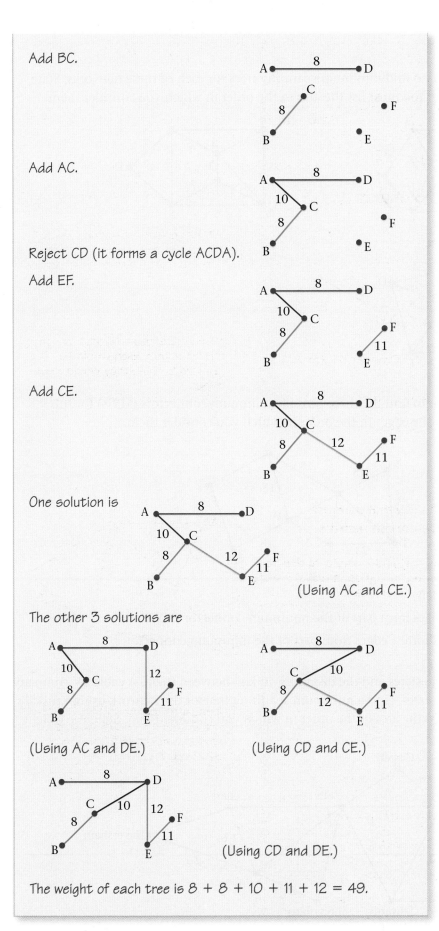

Notice that, when using Kruskal's algorithm, the chosen arcs seem to jump around the network. Remember Kruskal's algorithm seems *cha*otic.

Looking at the list of arcs, we can see that, although we included both arcs of length 8, we only included one arc of length 10 and one arc of length 12. We can never select both length 10 or both length 12 arcs. However, we must always choose one of each. This helps to determine the other three solutions.

Exercise 3A

1 Use Kruskal's algorithm to find minimum spanning trees for each of these networks. State the weight of each tree. You must list the arcs in the order in which you consider them.

a
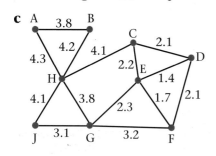

b
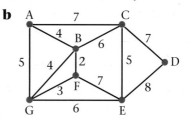

c

2 Use Kruskal's algorithm to find the three possible minimum connectors (MSTs) for this network. You must list the edges in the order in which you consider them.

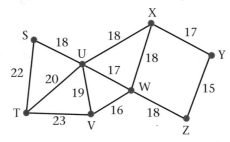

3 Draw a network in which

a the three shortest edges form part of the minimum connector (MST),

b not all of the three shortest edges from part of the minimum connector.

4 The diagram shows ten estates and the distances, in km, between them. A cable TV company wishes to link up the estates. Find a minimum spanning tree for the network using Kruskal's algorithm. You must list the arcs in the order in which you consider them. State the weight of your tree.

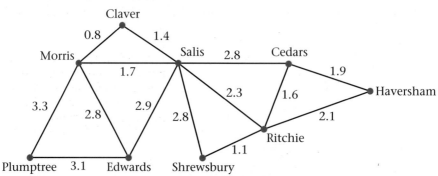

3.2 You can use Prim's algorithm on a network to find a minimum spanning tree.

■ Like Kruskal's algorithm, Prim's algorithm finds the minimum spanning tree, but it uses a different approach.

■ Here is Prim's algorithm.

1 Choose any vertex to start the tree.

2 • Select an arc of least weight that joins a vertex that is already in the tree to a vertex that is not yet in the tree.

  • If there is a choice of arcs of equal weight, choose randomly.

3 Repeat step 2 until all the vertices are connected.

Example 3

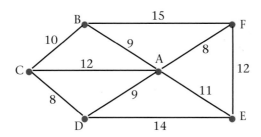

In the examination, you may be asked to use Prim's or Kruskal's algorithm. You must know which is which.

Use Prim's algorithm to find a minimum spanning tree for the network above. List the arcs in the order in which you add them to your tree.

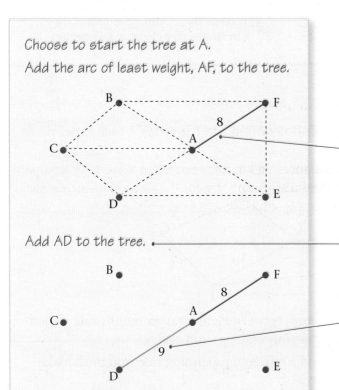

Choose to start the tree at A.

Add the arc of least weight, AF, to the tree.

The arcs we consider are those linking A to another vertex. AF(8), AB(9), AD(9), AE(11) and AC(12). Add the one of least weight, AF, to the tree.

Add AD to the tree.

The arcs linking A or F to other vertices are AB(9), AD(9), AE(11), AC(12), FE(12) and FB(15).

Add the least arc, from A *or* F, that introduces a new vertex to the tree. In this case there are two to choose from, AD or AB. We need to choose randomly, so we can choose AD.

Add DC to the tree.

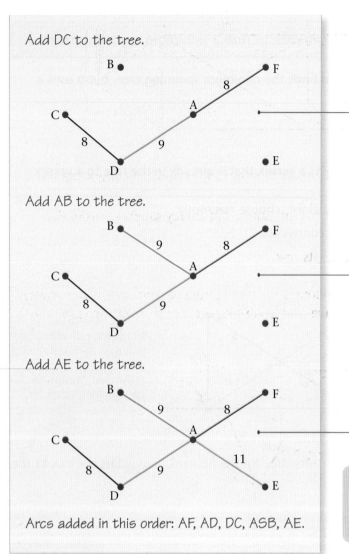

The arcs linking A, F and D to the remaining vertices are AB(9), AE(11), AC(12), FE(12), FB(15), DC(8), DE(14). The one of least weight is DC.

Add AB to the tree.

The arcs linking A, F, D and C to the remaining vertices (B and E) are AB(9), AE(11), FE(12), FB(15), DE(14), CB(10). The one of least weight is AB.

Add AE to the tree.

The arcs linking A, F, D, C and B to the remaining vertex, E, are AE(11), FE(12), DE(14). The one of least weight is AE.

Notice that with Prim's algorithm the tree *always* 'grows' in a connected fashion. The arcs do no 'jump around' as they sometimes do in Kruskal's algorithm.

Arcs added in this order: AF, AD, DC, ASB, AE.

Exercise 3B

1 Repeat Question **1** in Exercise 3A using Prim's algorithm. Start at vertex A each time.

2

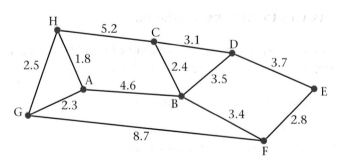

The network shows the distance, in kilometres, between eight weather monitoring stations. The eight stations need to be linked together with underground cables.

a Use Prim's algorithm, starting at A, to find a minimum spanning tree. You must make your order of arc selection clear.

b Given that cable costs £850 per kilometre to lay, find the cost of linking these weather stations.

3

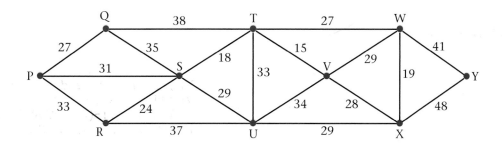

The network shows ten villages and the costs, in thousands of pounds, of connecting them with a new energy supply.

Use Prim's algorithm starting at P, to find the minimum cost energy supply network that would connect all ten villages.

Draw your minimum connector and state its cost.

4 Use Prim's algorithm, starting at A, to find four distinct minimum connectors for the network below. In each case draw your spanning tree, and make your order of arc selection clear.

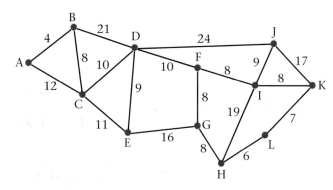

3.3 You can apply Prim's algorithm to a distance matrix.

- Networks (particularly large ones) are often described in distance matrix form.

- Networks can be inputted into computers in this form.

- Prim's algorithm (unlike Kruskal's algorithm) is easily adapted to a distance matrix, so it is Prim's algorithm, in matrix form, that is best suited to computerisation.

- Here is the distance matrix form of Prim's algorithm.

  1 Choose any vertex to start the tree.

  2 Delete the **row** in the matrix for the chosen vertex.

  3 Number the **column** in the matrix for the chosen vertex.

  4 Put a ring round the lowest undeleted entry in the numbered columns. (If there is an equal choice, choose randomly.)

  5 The ringed entry becomes the next arc to be added to the tree.

  6 Repeats steps 2, 3, 4 and 5 until all rows are deleted.

Example 4

	A	B	C	D	E
A	–	27	12	23	74
B	27	–	47	15	71
C	12	47	–	28	87
D	23	15	28	–	75
E	74	71	87	75	–

Apply Prim's algorithm to the distance matrix above to find a minimum spanning tree. Start at A.

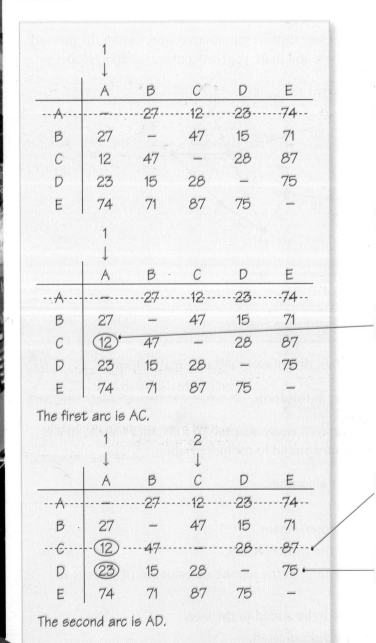

The first arc is AC.

The second arc is AD.

Delete row A and number column A.

We now seek the smallest entry in the A column – which is 12.

The lowest undeleted entry in column A is 12, so put a ring round it. The first arc is AC.

The new vertex is C. Delete the C row and number the column, C.

The lowest undeleted entry in columns A and C is 23, so put a ring round it. The second arc is AD.

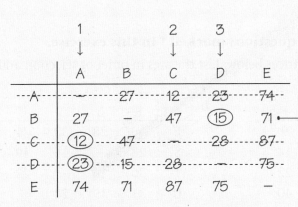

	1 ↓ A	2 ↓ B	3 ↓ C	D	E
~~A~~	—	27	12	23	74
B	27	—	47	(15)	71
~~C~~	(12)	47	—	28	87
~~D~~	(23)	15	28	—	75
E	74	71	87	75	—

The new vertex is D. Delete row D, and number column D.

The lowest undeleted entry in columns A, C and D is 15. Put a ring round it. The third arc is DB.

The third arc is DB.

	1 ↓ A	4 ↓ B	2 ↓ C	3 ↓ D	E
~~A~~	—	27	12	23	74
~~B~~	27	—	47	(15)	71
~~C~~	(12)	47	—	28	87
~~D~~	(23)	15	28	—	75
E	74	(71)	87	75	—

The new vertex is B. Delete row B, and number column B.

The lowest undeleted entry in columns A, C, D and B is 71. Put a ring round it. The fourth arc is BE.

The fourth arc is BE.

	1 ↓ A	4 ↓ B	2 ↓ C	3 ↓ D	5 ↓ E
~~A~~	—	27	12	23	74
~~B~~	27	—	47	(15)	71
~~C~~	(12)	47	—	28	87
~~D~~	(23)	15	28	—	75
~~E~~	74	(71)	87	75	—

The new vertex is E. Delete row E and number column E.

All rows have now been deleted, so the algorithm is complete.

> You do not need to show all of these tables. The final, labelled table, plus a list of arcs in order, is sufficient to make your method clear.

The minimum connector is

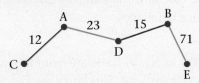

Its weight is 12 + 23 + 15 + 71 = 121

49

Exercise 3C

Photocopy masters are available for the questions marked * in this exercise.

1 * Apply Prim's algorithm to the distance matrices below. List the arcs in order of selection and state the weight of your tree.

a

	A	B	C	D	E	F
A	–	15	20	34	25	9
B	15	–	36	38	28	14
C	20	36	–	43	38	22
D	34	38	43	–	26	40
E	25	28	38	26	–	31
F	9	14	22	40	31	–

b

	R	S	T	U	V
R	–	28	30	31	41
S	28	–	16	19	43
T	30	16	–	22	41
U	31	19	22	–	37
V	41	43	41	37	–

2 *

	Birmingham	Nottingham	Lincoln	Stoke	Manchester
Birmingham	–	164	100	49	88
Nottingham	164	–	37	56	74
Lincoln	100	37	–	90	86
Stoke	49	56	90	–	44
Manchester	88	74	86	44	–

The table shows the distance, in miles, between 5 cities. It is intended to link these 5 cities by a transit system.

Use Prim's algorithm, starting at Birmingham, to find a minimum spanning tree for this network. You must list the arcs in order of selection and state the weight of your tree.

3 *

	A	B	C	D	E	F	G	H
A	–	84	53	35	–	47	–	42
B	84	–	71	113	142	61	75	–
C	53	71	–	–	–	–	59	–
D	35	113	–	–	58	67	151	–
E	–	142	–	58	–	168	159	48
F	47	61	–	67	168	–	–	73
G	–	75	59	151	159	–	–	52
H	42	–	–	–	48	73	52	–

The table shows the costs, in euros per 1000 words, of translating DVD player instruction manuals between eight languages.

a Use Prim's algorithm, starting from D, to find the cost of translating an instruction manual of 3000 words from D into the seven other languages.

b Draw your minimum spanning tree.

From the table we see that it costs 159 euros per 1000 words to translate from language E to G. A manual is written in language E and needs to be translated into language G.

c Give a reason why
   i it might be decided *not* to translate directly from E to G,
   ii it might be decided to translate directly.

4 *

	X	A	B	C	D	E	F	G	H	I
X	–	65	80	89	74	26	71	41	41	74
A	65	–	27	41	22	37	20	29	25	43
B	80	27	–	30	24	55	16	46	40	42
C	89	41	30	–	50	84	24	70	49	26
D	74	22	24	50	–	51	35	34	47	63
E	26	37	55	84	51	–	52	18	23	68
F	71	20	16	24	35	52	–	45	31	27
G	41	29	46	70	34	18	45	–	25	64
H	41	25	40	49	47	23	31	25	–	44
I	74	43	42	26	63	68	27	64	44	–

The table shows the distances, in miles, between nine oil rigs and the depot X. Pipes are to be laid to connect the rigs and the depot.

a Use Prim's algorithm, starting at X, to find a minimum connector for the network. You must make the order of arc selection clear.

Oil rig A exhausts its supply and is closed down.

b Use Prim's algorithm to find a minimum connector excluding A. You must make the order of arc selection clear.

3.4 You can use Dijkstra's algorithm to find the shortest path between two vertices in a network.

■ Dijkstra's algorithm is used to find the shortest, cheapest or quickest route between two vertices. An example could be finding the shortest cycle route from John O'Groats to Lands End.

> Dijkstra is pronounced Dike-stra.

■ Here is Dijkstra's algorithm
(to find the shortest path from S to T through a network).

1 Label the start vertex, S, with the final label, 0.

2 Record a working value at *every* vertex, Y, that is directly connected to the vertex, X, that has just received its final label.
- Working value at Y = final value at X + weight of arc XY
- If there is already a working value at Y, it is only replaced if the new value is smaller.
- Once a vertex has a final label it is not revisited and its working values are no longer considered.

3 Look at the working values at all vertices without final labels. Select the smallest working value. This now becomes the final label at that vertex. (If two vertices have the same smallest working value either may be given its final label first.)

4 Repeat steps 2 and 3 until the destination vertex, T, receives its final label.

5 To find the shortest path, trace back from T to S. Given that B already lies on the route, include arc AB whenever final label of B − final label of A = weight of arc AB.

> The algorithm makes use of labels. Start at the initial vertex and move through the network, putting *working values* (often called temporary labels) on each vertex. Each pass finds the shortest route to one of the vertices and records its *final label* (also called its permanent label). Once a vertex has its final label it is 'sealed' and its working values are no longer considered. Continue in this way until the destination vertex is reached.

Example 5

Use Dijkstra's algorithm to find the shortest route from S to T in the network below.

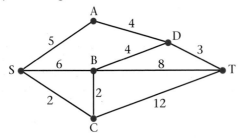

To make the working clear we replace the vertices with boxes like this

Vertex	Order of labelling	Final value
Working values		

In the examination the boxes will be drawn for you to complete.

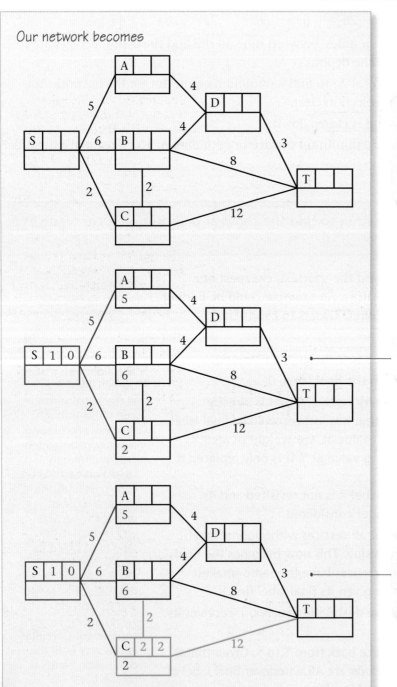

Our network becomes

Give vertex S a final label of 0. Indicate that this is the first vertex to receive its final label by completing the top boxes at S like this.

S	1	0

Give working values to A, B and C since they are *directly* connected to S.

Look at the working values at A, B and C. The smallest is 2 (at C). This will become the next final label. C is now completed. It is the 2nd vertex to be completed.

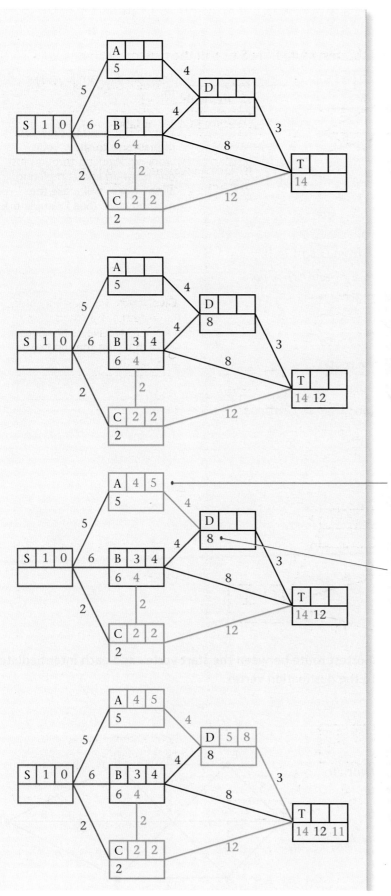

Add working values to each vertex that is directly connected to C. [Note that the algorithm has picked up the short cut to B, S → B is 6, but S → C → B is 4.]

The smallest working value is 4 at B. This becomes the final label at B. B becomes the 3rd completed vertex.
Add working values to D and T because these are directly connected to B.

The smallest working value is the 5 at A. A becomes the 4th vertex to be completed. It has a final value of 5.

The only working value to add would be a 9 at D. However this is larger than the 8 already there, so it does not need to be recorded.

The smallest working value is the 8 at D. This becomes D's final label. The only working value to add is the 11 at T. D becomes the 5th vertex to be completed.

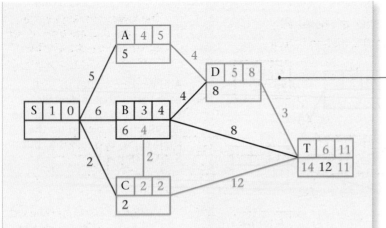

The final diagram looks like this.

You do not need to draw lots of diagrams to show your working. Working through just one diagram and completing all the boxes will make the method clear. (See Example 6.)

The length of the shortest route from S to T is 11.

To find the shortest route, start at T and trace back looking at final values and arc lengths.

Check the arcs into T.

    11 − 2 ≠ 12 so CT is not on the route.

    11 − 4 ≠ 8 so BT is not on the route.

    11 − 8 = 3 so DT is on the route.

To get to T in 11 we must have come from D. Continue working back from D.

The working is

    11 − 8 = 3   DT

    8 − 4 = 4   BD

    4 − 2 = 2   CB

    2 − 0 = 2   SC

so the shortest route is

S C B D T, length 11.

■ **Dijkstra's algorithm finds the shortest route between the start vertex and each intermediate vertex completed on the way to the destination vertex.**

Example 6

Use Dijkstra's algorithm in this network to find the length of the shortest route

a from A to H,

b from A to G.

List the routes you use.

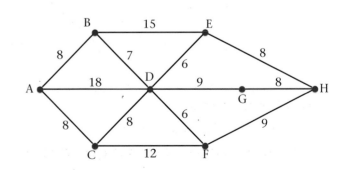

a The final diagram looks like this. It shows all
the working needed to make the method clear.

> The lowest working values
> at B and at C are both 8, so
> either can be chosen first.

```
      B 3 8        15        E 6 21
      8                      23 21
    8            7         6              8
 A 1 0      18      D 4 15       9      G 7 24    8    H 8 29
                   18 16 15              24           29
    8            8         6
      C 2 8                    F 5 20          9
      8            12          20
```

The length of the shortest route from A to H is 29.

There are two shortest routes:

A B D E H and A C F H.

b From the diagram above the shortest route from
A to G is 24.

The route is A B D G.

$$24 - 15 = 9 \quad DG$$
$$15 - 8 = 7 \quad BD$$
$$8 - 0 = 8 \quad AB$$

> either
> $29 - 21 = 8$ EH
> $21 - 15 = 6$ DE
> $15 - 8 = 7$ BD
> $8 - 0 = 8$ AB
> or
> $29 - 20 = 9$ FH
> $20 - 8 = 12$ CF
> $8 - 0 = 8$ AC

> Check that your method
> can be followed clearly.

■ It is possible to use Dijkstra's algorithm on networks with directed arcs.
This is like trying to find a route where some of the roads are one-way streets.

Example 7

The network represents part of a road system in a city. Some roads are one-way and these are
indicated by directed arcs. The number on each arc represents the time, in minutes, to travel
along that arc.

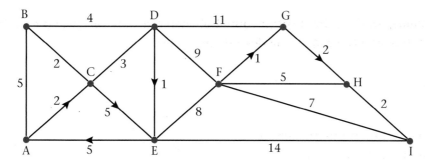

a Show that there are two quickest routes from A to I. Explain how you found your routes from
your labelled diagram.

Road HI is closed due to roadworks.

b Find the quickest route from A to I, avoiding HI.

a The final diagram looks like this.

> Check that your method can be understood from this diagram.

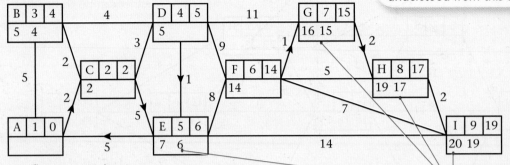

The two quickest routes are

A C D F G H I and A C D E F G H I

since		
19 − 17 = 2	HI	
17 − 15 = 2	GH	
15 − 14 = 1	FG	
14 − 5 = 9	DF	
5 − 2 = 3	CD	
2 − 0 = 2	AC	

since
19 − 17 = 2 HI
17 − 15 = 2 GH
15 − 14 = 1 FG
14 − 6 = 8 EF
6 − 5 = 1 DE
5 − 2 = 3 CD
2 − 0 = 2 AC

> Examiners check the *order* in which the numbers appear in the list of working values. It is important to put them in the order given by the algorithm.

Both are of length 19 minutes.

b Removing HI from the network would leave a final value of 20 at I. Start at I and find the route of length 20.

20 − 6 = 14 EI
6 − 5 = 1 DE
5 − 2 = 3 CD
2 − 0 = 2 AC

so the quickest route from A to I, avoiding HI, is A C D E I, length 20 minutes.

Exercise 3D

Photocopy masters are available for the questions marked * in this exercise.

1 *Use Dijkstra's algorithm to find a shortest route from S to T in each of the following networks. Show your working. State your shortest routes and their lengths. You should show how you obtained your shortest route from your labelled diagrams.

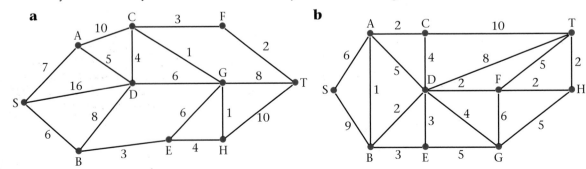

c

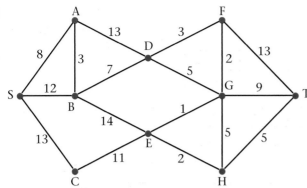

2 * The network shows part of a road
network in a city. The number on each
arc gives the travel time, in minutes, it
takes to travel along that arc.

Find

a the quickest route from A to Q
and its length,

b the quickest route from A to L
and its length,

c the quickest route from M to A and its length,

d the quickest route from P to A and its length.

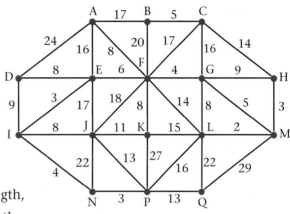

3 * Use Dijkstra's algorithm to
find the shortest route, and
its length, from A to F in the
directed network opposite.

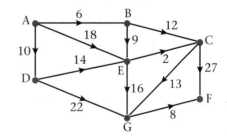

4 * The network represents the distances, in metres, of all the roads in a building site. A crane
is needed for one day at T. There are two cranes available on site, one at S_1, and the other at
S_2. One of these two cranes will be moved to T. In order to minimise the cost it is decided to
move the crane that is closest to T. Use Dijkstra's algorithm to determine which crane should
be moved.

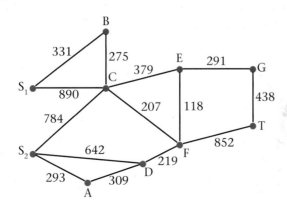

It is possible to solve this
problem with only one
application of Dijkstra's
algorithm. Think carefully
about the starting point.

5 *

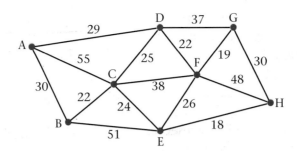

a Use Dijkstra's algorithm to find the shortest route from A to H. Indicate how you obtained your shortest route from your labelled diagram.

b Find the shortest route from A to H via G.

c Find the shortest route from A to H, not using CE.

6 *

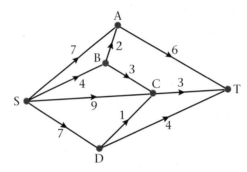

Use Dijkstra's algorithm to find the shortest route from S to T. State the length of your route.

Mixed exercise 3E

Photocopy masters are available for questions marked * in this exercise.

1 The network represents a theme park with seven zones. The number on each arc shows a distance in metres.

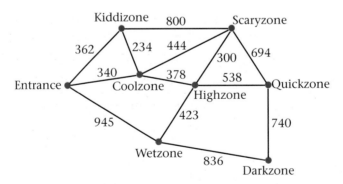

Tramways are to be built to link the seven zones and the car park at the Entrance.

a Find a minimum connector using

  i Kruskal's algorithm,

  ii Prim's algorithm, starting at the Entrance.

You must make your order of arc selection clear.

b Draw your tree and state its weight.

2 The network represents eight observation points in a wildlife reserve and the possible paths connecting them. The number on each arc is the distance, in kilometres, along that path. It is decided to link the observation points by paths, but in order to minimise the impact on the wildlife reserve, we wish to use the least total length of path.

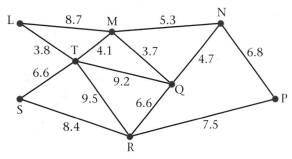

a Find a minimum spanning tree for the network using

   i   Prim's algorithm, starting at L,   **ii**   Kruskal's algorithm.

   In each case list the arcs in the order in which you consider them.

Given that paths TQ and RP already exist and so will form part of the tree,

b state which algorithm, Prim's or Kruskal's, you would select to complete the spanning tree. Give a reason for your answer.

3 *

	A	B	C	D	E	F
A	–	124	52	87	58	97
B	124	–	114	111	115	84
C	52	114	–	67	103	98
D	87	111	67	–	41	117
E	58	115	103	41	–	121
F	97	84	98	117	121	–

The table shows the distances, in mm, between six nodes A to F in a network.

a Use Prim's algorithm, starting at A, to solve the minimum connector problem for this table of distances. You must explain your method carefully and indicate clearly the order in which you selected the arcs.

b Draw a sketch showing the minimum spanning tree and find its length.   **E**

4 *It is intended to network five computers at a large theme park. There is one computer at the office and one at each of the four different entrances. Cables need to be laid to link the computers. Cable laying is expensive, so a minimum total length of cable is required.

The table shows the shortest distances, in metres, between the various sites.

	Office	Entrance 1	Entrance 2	Entrance 3	Entrance 4
Office	–	1514	488	980	945
Entrance 1	1514	–	1724	2446	2125
Entrance 2	488	1724	–	884	587
Entrance 3	980	2446	884	–	523
Entrance 4	945	2125	587	523	–

a Starting at Entrance 2, demonstrate the use of Prim's algorithm and hence find a minimum spanning tree. You must make your method clear, indicating the order in which you selected the arcs in your final tree.

b Calculate the minimum total length of cable required.   **E**

5

You are to use Kruskal's algorithm to find a minimum spanning tree for the network shown.

a i Write down the order in which you selected the arcs.

ii Sketch your minimum spanning tree.

iii State the weight of your minimum spanning tree.

For any connected network,

E = the number of edges in the minimum spanning tree, and

V = the number of vertices in the network.

b Write down the relationship between E and V.

E

6 A company is to install power lines to buildings on a large industrial estate. The lines are to be laid by the side of the roads on the estate. The estate is shown as a network opposite. The buildings are designated A, B, C, ..., N and the distances between them are given in hundreds of metres. The manager wants to minimise the total length of power line to be used.

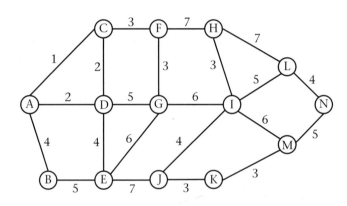

a Use Kruskal's algorithm to obtain a minimum spanning tree for the network and hence determine the minimum length of power line needed.

Owing to a change of circumstances, the company modifies its plans for the estate. The result is that the road from F to G now has a length of 700 metres.

b Determine the new minimum total length of power line.

7 *

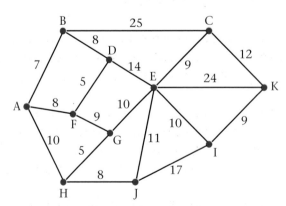

A weighted network is shown above. The number on each arc indicates the weight of that arc.

a Use Dijkstra's algorithm to find a path of least weight from A to K.

State clearly

   i   the order in which the vertices were labelled,

   ii how you determined the path of least weight from your labelling.

b List all alternative paths of least weight.

c Describe a practical problem that could be modelled by the above network and solved using Dijkstra's algorithm.

8 *

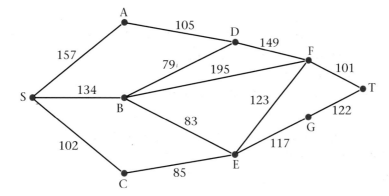

The network above shows the distances, in miles, between nine cities. Use Dijkstra's algorithm to determine the shortest route, and its length, between cities S and T. You must indicate clearly

   i   the order in which the vertices are labelled,

   ii how you used your labelled diagram to decide which cities to include in the shortest route.

Summary of key points

1   Kruskal's algorithm and Prim's algorithm can be used to find a minimum spanning tree, but they take different approaches.

2   To use Kruskal's algorithm, you sort the arcs into ascending order of weight and use the arc of least weight to start the tree. You then add arcs in order of ascending weight, unless an arc would form a cycle, in which case it is rejected.

3   To use Prim's algorithm, you choose any vertex to start the tree. You then select an arc of least weight that joins a vertex that is already in the tree to a vertex that is not yet in the tree. You repeat this until all vertices are connected.

4   Prim's algorithm can be applied to a distance matrix. You choose any vertex to start the tree. You delete the row in the matrix for the chosen vertex and number the column in the matrix for the chosen vertex. You ring the lowest undeleted entry in the labelled columns, which becomes the next vertex. You repeat this until all rows are deleted.

5   You can use Dijkstra's algorithm to find the shortest path between two vertices in a network.

6 The start vertex is given the final value 0. Every vertex that is directly connected to the start vertex is given a working value

final value at start + weight of arc.

Select the smallest working value. This is the final value for that vertex. Repeat until the destination vertex receives its final label. Find the shortest path by tracing back from destination to start. Include an arc AB on the route if B is already on the route and

final label B − final label A = weight of arc AB

7 Each vertex is replaced by a box

Vertex	Order of labelling	Final value
Working values		

After completing this chapter you should be able to:

1 determine whether a graph is traversable

2 use the route inspection (Chinese postman) algorithm to find the shortest inspection cycle in a network.

Route inspection (Chinese postman problem)

4

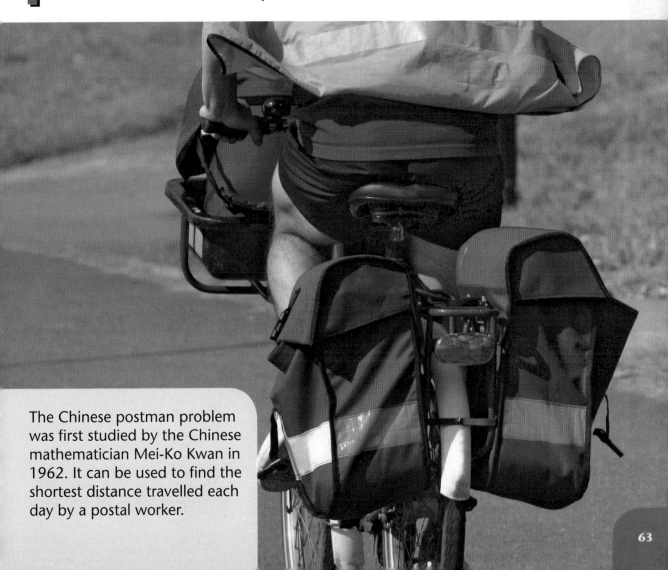

The Chinese postman problem was first studied by the Chinese mathematician Mei-Ko Kwan in 1962. It can be used to find the shortest distance travelled each day by a postal worker.

4.1 You can determine whether a graph is traversable.

■ In Chapter 2, you learnt that the **degree** or **valency** or **order** of a **vertex** is the **number of arcs incident to it**. A vertex is **odd** (**even**) if it has **odd** (**even**) valency.

■ If *all* the valencies in a graph are **even**, then the graph is **Eulerian**.

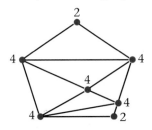

Each vertex has an even valency.

■ If precisely two valencies are odd, and all the rest are even, then the graph is semi-Eulerian.

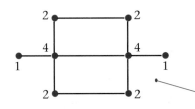

It is not possible to draw a graph with just one odd valency. (See Example 3.)

Precisely 2 vertices have odd valency.

■ A graph is **traversable** if it is possible to traverse (travel along) every arc just once without taking your pen from the paper.

■ A graph is traversable if all the valencies are even.

■ A graph is semi-traversable if it has precisely two odd valencies. In this case the start point and the finish point will be the two vertices with odd valencies.

■ A graph is not traversable if it has more than two odd valencies.

Example **1**

a Verify that the graph is Eulerian.

b Find a route, starting and finishing at A, that traverses the graph.

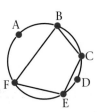

a

Vertex	A	B	C	D	E	F
Valency	2	4	4	2	4	4

All valencies are even, so the graph is Eulerian.

b A possible route is

A B C D E F B C E F A.

There are many other routes. All of them will be 11 letters long.

Example 2

Find a route that traverses each arc of this graph just once.
You may start and finish at different points.

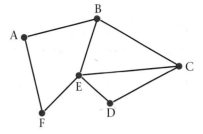

Vertex	A	B	C	D	E	F
Valency	2	3	3	2	4	2

This has precisely two odd valencies, so it is semi-Eulerian.

A possible route is

B A F E D C E B C.

The odd valencies are at B and C, so start at one of these and finish at the other.

Example 3

Prove that there must always be an even (or zero) number of vertices with odd valency in every graph.

This is called Euler's hand-shaking lemma.

Each arc has two ends and so will contribute two to the sum of the valencies of the whole graph.

⇒ The sum of the valencies = 2 × number of arcs

⇒ The sum of the valencies is even.

⇒ Any odd numbers must occur in pairs.

⇒ There is an even number of odd valencies.

This means that, in any graph, there will be zero, or two, or four, or six, or eight, ... vertices with odd valency.

Exercise 4A

1 List the valency of each vertex and hence determine if each of the graphs below are

  i Eulerian   **ii** semi-Eulerian   **iii** neither.

For those that are Eulerian or semi-Eulerian, find a route that traverses each arc just once.

a
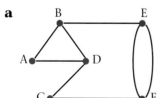

b
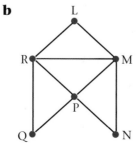

c
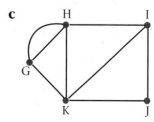

2 **a** Show that each of the graphs below is Eulerian.

b In each case, find a route that starts and finishes at A and traverses each arc just once.

i

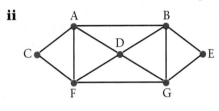

ii

3 **i** R   S

T   U

V   W

ii

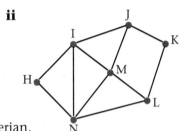

a Show that each of these graphs is semi-Eulerian.

b In each case, find a route, starting and finishing at different vertices, that traverses each edge just once.

4 Explain why each of the graphs below is not traversable.

a   B
A   C

D   E   F

G   I
H

b

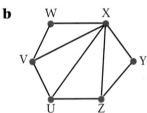

5 Considering the valencies of each vertex in Questions **1** to **4**, verify the hand-shaking lemma.

6 Explain why a traversable graph has either

a all of its vertices with even valency or

b precisely two vertices of odd valency, these being the start and finish points.

7 The diagram represents the city of Königsberg (Prussia, now Kaliningrad, Russia). The Pregel river runs through the city and creates two large islands in the centre. The two islands (C and D) were linked to each other and the mainland (A and B) by seven bridges.

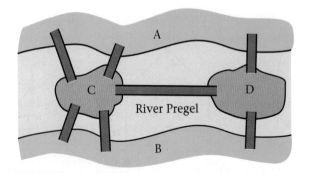

The problem for the citizens of Königsberg was to decide whether or not it was possible to walk a route that crossed each bridge just once and returned to its starting point.

> This is the famous 'Bridges of Königsberg' problem.

a Using four vertices A, B, C and D to represent the four parts of the city, and seven arcs to represent the bridges, draw a graph to model the problem.

b Show that the graph is not traversable.

There is a continuation of the problem. Johannes works at A, Gregor works at B and Peter works at D. There is a hotel at C.

c Johannes builds an eighth bridge so that he can start at A and finish at his home at C, crossing each bridge once. However, he does not want Gregor to be able to find a similar route from B to C. Where should Johannes build his eighth bridge?

d Gregor decides to build a ninth bridge so that he can start at B and finish at his home near C, crossing each bridge once. He does not want Johannes to be able to find a similar route from A to C. Where should Gregor build his ninth bridge?

e Peter decides to build a tenth bridge, so that every person in the city can cross all the bridges in turn and return to their starting point. Where should Peter build the tenth bridge?

4.2 You can use the route inspection (Chinese postman) algorithm to find the shortest route in a network.

■ This algorithm can be used to find the shortest route that traverses every arc at least once and **returns to the starting point**.

■ If *all* the vertices have even valency the network is traversable. The length of the shortest route will be equal to the weight of the network.

> In general there will be many different shortest inspection routes, but all will have the same minimum length.

Example 4

Solve the route inspection problem for the network below, starting and finishing at A.

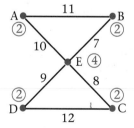

All valencies are even, so the network is traversable. A possible route is

A B E C D E A

length 11 + 7 + 8 + 12 + 9 + 10 = 57

There are several other solutions such as A B E D C E A or A E D C E B A but all will have length 57.

■ If there are only two odd vertices, repeat the shortest path between them, and add it to the network.

Example 5

Find a minimal route, starting and finishing at S, that traverses each edge at least once.

State your route and its length.

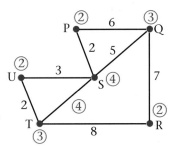

Weight of network
$= 5 + 4 + 8 + 7 + 6 + 2 + 3 + 2 = 37$

The odd valencies are at vertices Q and T. We have to repeat the shortest path Q to T, Q S T of length 9.

We add QS and ST to the diagram so it becomes

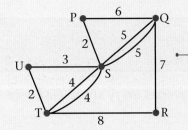

Adding the extra arcs now makes the network Eulerian and so it can be traversed.

A possible route is

S Q P S Q R T S U T S

length = 37 (weight of network) + 9 (length of repeats)
= 46

In the examination the shortest route can be found by inspection. If Dijkstra's algorithm is required, you will be directed to use it, possibly earlier in the question.

For this example, all correct routes will have a length of 46 and will be 11 letters long.

■ If there are more than two odd valencies, you need to consider all possible complete pairings and select the one that gives the smallest total and then add this pairing to the network.

You will only have to consider at most four odd valencies in the examination.

Example 6

Solve the Chinese postman problem for the network below, starting and finishing at A.

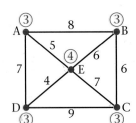

Weight of network
$= 8 + 6 + 9 + 7 + 5 + 6 + 4 + 7$
$= 52$

A, B, C and D have odd valency.

You can pair all these four nodes in three ways

A with B and C with D

A with C and B with D

A with D and B with C.

These give the following lengths.

AB + CD = 8 + 9 = 17

AC + BD = 12 + 10 = 22

AD + BC = 7 + 6 = 13 ← least sum

Add AD and BC to the network.

> In the examination you *must* consider all three pairings of the four odd nodes.

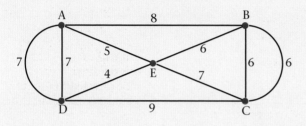

A possible route is

    A B C E B C D E A D A

length 52 + 13 = 65

> For this example all shortest routes will be of length 65 and be 11 letters long.

■ Here is the route inspection algorithm.

  1 Identify any vertices with odd valency.

  2 Consider all possible complete pairings of these vertices.

  3 Select the complete pairing that has the least sum.

  4 Add a repeat of the arcs indicated by this pairing to the network.

Example 7

a Solve the route inspection problem for this network starting and finishing at A.

Given that it is now permitted to start and finish at different vertices

b Select the start and finish vertices that give the shortest route, explaining your reasoning. State the length of your route.

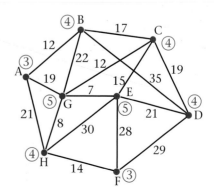

a There are odd valencies, at A E F and G.

By inspection path lengths are:

AE + FG = 26 + 22 = 48

AF + EG = 35 + 7 = 42 ← least sum

AG + EF = 19 + 28 = 47

We need to repeat arcs AH, HF and EG and add them to the network.

The weight of the network is 309.

The length of the shortest route will be 309 + 42 = 351

Modify the network by adding the extra arcs AH, HF and EG.

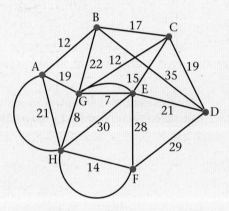

A possible route is
A B C D B G C E G A H E D F E G H F H A.

b One vertex with odd valency as the start vertex and a second as the finish vertex.

That leaves just one pair of vertices with odd valency.

Repeat the path between them.

Choose the pair which has the shortest path between them.

Looking at the figures at the start of **a**, the smallest of the six numbers is 7, which corresponds to EG.

Choose to repeat EG, so use A and F as the start and finishing vertices.

The length of the new route is
309 + 7 = 316

Shortest paths as follows
AE is AGE = 26
FG is FHG = 22
AF is AHF = 35

The shortest path AF is AHF.

The weight of the network is found by summing the weights of the arcs.

You were not asked to find the route, so you don't need to list it.

Exercise 4B

1 Solve the route inspection problem for each of the networks below. In each case, state your minimal route and its length.

a

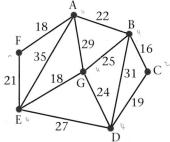

b

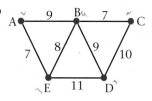

c

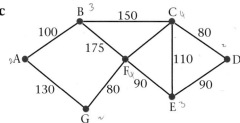

d

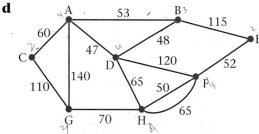

2 Each of the diagrams below show a network of roads that need to be inspected. In each case, find the length of the shortest route that traverses each arc at least once and returns to the start vertex. State your routes.

a

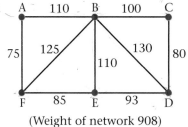

(Weight of network 908)

b

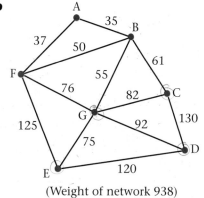

(Weight of network 938)

c

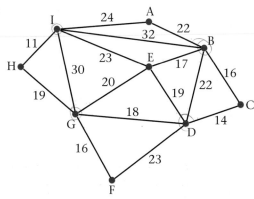

(Weight of network 326)

3 The diagram shows the paths in a park. The number on each arc gives the length, in metres, of that path. The vertices show the park entrances, A, B, C, D, E and F.

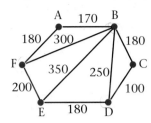

A gardener needs to inspect each path for weeds.

She will walk along each path once and wishes to minimise her route.

a Use the route inspection algorithm to find a minimum route, starting and finishing at entrance A. State the length of your route.

Given that it is now permitted to start and finish at two different entrances,

b find the start and finish points that would give the shortest route, and state the length of the route.

4 The diagram represents a system of roads.

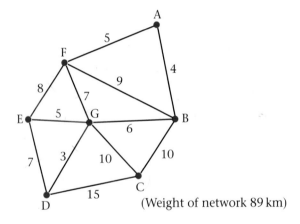

(Weight of network 89 km)

The number on each arc gives the distance, in kilometres, of that road.

The town council needs to renew the road markings.

Cherry will be renewing the kerbside markings and Mac will renew the centre road markings.

Cherry needs to travel along each road twice, once on each side of the road.

a Explain how this differs from the standard route inspection problem and find the length of Cherry's route.

Mac must travel along each road once.

b Use the route inspection algorithm to find a minimal route. You should state the roads he will traverse twice and the length of his route.

Road EG is being resurfaced soon and it is decided not to renew its road markings until after the resurfacing.

Given that EG may be omitted from his route,

c find the length of Mac's minimal route.

Mixed exercise 4C

1 The network of paths in a garden is shown below. The numbers on the paths give their lengths in metres. The gardener wishes to inspect each of the paths to check for broken paving slabs so that they can be repaired before the garden is opened to the public. The gardener has to walk along each of the paths at least once.

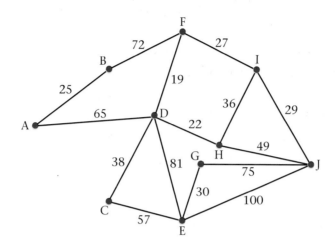

a Write down the degree (valency) of each of the ten vertices A, B, …, J.

b Hence find a route of minimum length. You should clearly state, with reasons, which, if any, paths will be covered twice.

c State the total length of your shortest route.

E

2

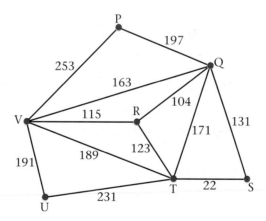

Starting and finishing at P, solve the route inspection (Chinese postman) problem for the network shown above. You must make your method and working clear.

State:

a your route, using vertices to describe the arcs,

b the total length of your route.

E

3

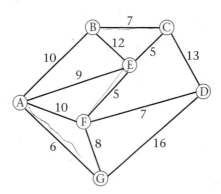

The diagram shows the network of paths in a garden to be opened to the public. The number on each path gives its length in metres. The gardener wishes to inspect each of the paths to check for broken paving slabs, so that they can be repaired before the garden is opened.

a Write down the degree (valency) of the seven vertices A, B, C, D, E, F and G.

b Use an appropriate algorithm to find a route of minimum length which starts and finishes at A and which traverses each path at least once. Write down which paths, if any, will be traversed twice.

c Calculate the total length of your shortest route.

E

4 The network shows the major roads that are to be gritted by a council in bad weather. The number on each arc is the length of the road in kilometres.

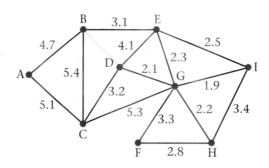

a List the valency of each of the vertices.

b Starting and finishing at A, use an algorithm to find a route of minimum length that covers each road at least once. You should clearly state, with reasons, which (if any) roads will be traversed twice.

c Obtain the total length of your shortest route.

There is a minor road BD (not shown) between B and D of length 6.4 km. It is not a major road so it does not need gritting urgently.

d Decide whether or not it is sensible to include BD as a part of the main gritting route, giving your reasons. (You may ignore the cost of the grit.)

E

5 The network opposite represents the streets in a village. The number on each arc represents the length of the street in metres.

The junctions have been labelled A, B, C, D, E, F, G and H.

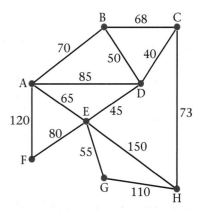

An aerial photographer has taken photographs of the houses in the village. A salesman visits each house to see if the occupants would like to buy a photograph of their house. He needs to travel along each street at least once. He parks his car at A and starts and finishes there. He wishes to minimise the total distance he has to walk.

a Describe an appropriate algorithm that can be used to find the minimum distance the salesman needs to walk.

b Apply the algorithm and hence find a route that the salesman could take, stating the total distance he has to walk.

c A friend offers to drive the salesman to B at the start of the day and collect him from C later in the day.

Explaining your reasoning, carefully determine whether this would increase or decrease the total distance the salesman has to walk.

E

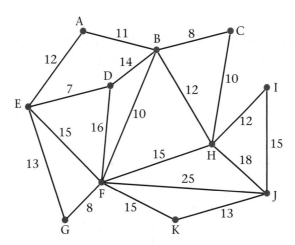

a Describe an algorithm that is used to solve the route inspection (Chinese postman) problem.

b Apply the algorithm and find a route, starting and finishing at A, that solves the route inspection problem for the network shown.

c State the total length of your route.

The situation is now altered so that, instead of starting and finishing at A, the route starts at one vertex and finishes at another vertex.

d i State the starting vertex and the finishing vertex which minimises the total length of the route. Give a reason for your selections.

ii State the length of your route.

e Explain why, in any network, there is always an even number of vertices of odd degree.

E

Summary of key points

1 A graph is traversable if it is possible to traverse (travel along) every arc just once without taking your pen from the paper.

2 A graph is traversable if all the valencies are even.

3 A graph is semi-traversable if it has precisely two odd valencies. In this case, the start and finish point will be the two vertices with odd valencies.

4 A graph is not traversable if it has more than two odd valencies.

5 The route inspection algorithm can be used to find the shortest route that traverses every arc at least once and **returns to the starting point**.

6 If there are only two odd vertices, repeat the shortest path between them and add it to the network.

7 If there are more than two odd valencies, consider all possible complete pairings, select the one that gives the smallest total, then add this pairing to the network.

Review Exercise

Photocopy masters are available for the questions marked * in this exercise.

1 The table opposite shows the points obtained by each of the teams in a football league after they had each played 6 games. The teams are listed in alphabetical order. Carry out a quick sort to produce a list of teams in descending order of points obtained.

Ashford	6
Colnbrook	1
Datchet	18
Feltham	12
Halliford	9
Laleham	0
Poyle	5
Staines	13
Wraysbury	14

E

A local council is responsible for maintaining pavements in a district. The roads for which it is responsible are represented by arcs in the diagram. The road junctions are labelled A, B, C, ..., G. The number on each arc represents the length of that road in km.

The council has received a number of complaints about the condition of the pavements. In order to inspect the pavements, a council employee needs to walk along each road twice (once on each side of the road) starting and ending at the council offices at C. The length of the route is to be minimal. Ignore the widths of the roads.

a Explain how this situation differs from the standard route inspection problem.

b Find a route of minimum length and state its length.

E

2

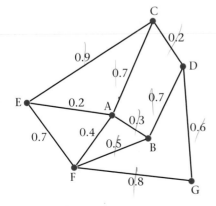

3 a Use the binary search algorithm to try to locate the name *SABINE* in the following alphabetical list. Explain each step of the algorithm.

    1 *ABLE*
    2 *BROWN*
    3 *COOKE*
    4 *DANIEL*
    5 *DOUBLE*
    6 *FEW*
    7 *OSBORNE*
    8 *PAUL*
    9 *SWIFT*
    10 *TURNER*

b Find the maximum number of iterations of the binary search algorithm needed to locate a name in a list of 1000 names. **E**

4 *

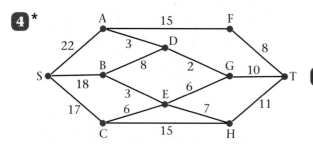

The diagram shows a network of roads. The number on each edge gives the time, in minutes, to travel along that road. Avinash wishes to travel from S to T as quickly as possible.

a Use Dijkstra's algorithm to find the shortest time to travel from S to T.

b Find a route for Avinash to travel from S to T in the shortest time. State, with a reason, whether this route is a unique solution.

On a particular day Avinash must include C in his route.

c Find a route of minimal time from S to T that includes C, and state its time. **E**

5 a State briefly
    i Prim's algorithm,
    ii Kruskal's algorithm.

b Find a minimum spanning tree for the network below using
    i Prim's algorithm, starting with vertex G,
    ii Kruskal's algorithm.

In each case write down the order in which you made your selection of arcs.

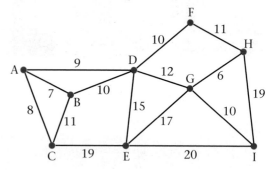

c State the weight of a minimum spanning tree.

d State, giving a reason for your answer, which algorithm is preferable for a large network. **E**

6

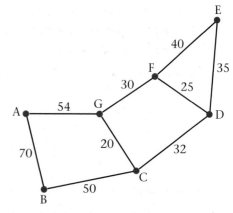

The diagram shows 7 locations A, B, C, D, E, F and G which are to be connected by pipelines. The arcs show the possible routes. The number on each arc gives the cost, in thousands of pounds, of laying that particular section.

a Use Kruskal's algorithm to obtain a minimum spanning tree for the network, giving the order in which you selected the arcs.

b Draw your minimum spanning tree and find the least cost of pipelines. **E**

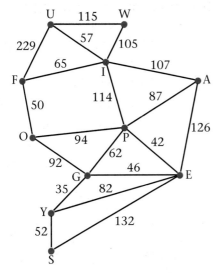

The network above represents the distances, in miles between eleven places A, E, F, G, I, O, P, S, U, W and Y.

a Use Dijkstra's algorithm to find the shortest route from W to S. State clearly
  i the order in which you labelled the vertices,
  ii how you determined the shortest route from your labelling,
  iii the places on the shortest route,
  iv the shortest distance.

b Explain how part **a** could have been completed so that the distance from A to S could also have been obtained without further calculation. (You are not required to find this distance.) **E**

8 * The network shows the possible routes between cities A, B, C, D, E, F, G and H.

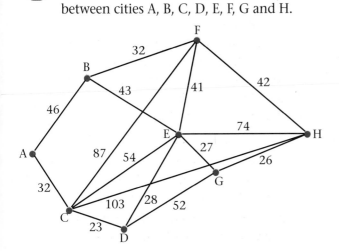

The number on each arc gives the cost, in pounds, of taking that part of the route. Use Dijkstra's algorithm to determine the cheapest route from A to H and its cost. Your solution must indicate clearly how you have applied the algorithm.

State clearly

a the order in which the vertices are labelled,

b how you used your labelled diagram to decide on the cheapest route. **E**

9 *

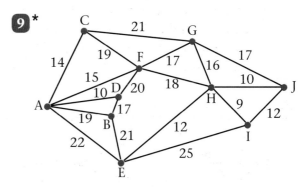

The network above models the roads linking ten towns A, B, C, D, E, F, G, H, I and J. The number on each arc is the journey time in minutes, along the road.

Alice lives in town A and works in town J.

a Use Dijkstra's algorithm to find the quickest route for Alice to travel to work each morning.

State clearly

  i the order in which all the vertices were labelled,

  ii how you determined the quickest route from your labelling.

b On her return journey from work one day Alice wishes to call in at the supermarket located in town C. Explain briefly how you would find the quickest route in this case. **E**

10 a Explain why it is impossible to draw a network with exactly three odd vertices.

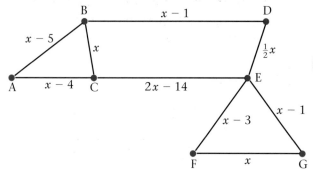

The route inspection problem is solved for the network above and the length of the route is found to be 100.

b Determine the value of x, showing your working clearly. **E**

11 a

	A	B	C	D	E	F
A	–	10	12	13	20	9
B	10	–	7	15	11	7
C	12	7	–	11	18	3
D	13	15	11	–	27	8
E	20	11	18	27	–	18
F	9	7	3	8	18	–

The table shows the distances, in metres, between six nodes A, B, C, D, E and F of a network.

  i Use Prim's algorithm, starting at A, to solve the minimum connector problem for this table of distances. Explain your method and indicate the order in which you selected the edges.

  ii Draw your minimum spanning tree and find its total length.

  iii State whether your minimum spanning tree is unique. Justify your answer.

b A connected network N has seven vertices.

  i State the number of edges in a minimum spanning tree for N.

A minimum spanning tree for a connected network has n edges.

  ii State the number of vertices in the network. **E**

12 a Describe the differences between Prim's algorithm and Kruskal's algorithm for finding a minimum connector of a network.

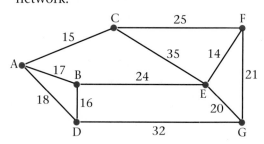

b Listing the arcs in the order that you select them, find a minimum connector for the network above, using

  i Prim's algorithm,

  ii Kruskal's algorithm. **E**

13   45 56 37 79 46 18 90 81 51

a Using the quick sort algorithm, perform one complete iteration towards sorting these numbers into ascending order.

b Using the bubble sort algorithm, perform one complete pass towards sorting the original list into descending order.

Another list of numbers, in ascending order, is

    7 23 31 37 41 44 50 62 71 73 94

c Use the binary search algorithm to locate the number 73 in this list. **E**

14 The following list gives the names of some students who have represented Britain in the International Mathematics Olympiad.

Roper (R), Palmer (P), Boase (B), Young (Y), Thomas (T), Kenney (K), Morris (M), Halliwell (H), Wicker (W), Garesalingam (G).

a Use the quick sort algorithm to sort the names above into alphabetical order.

b Use the binary search algorithm to locate the name Kenney. *E*

15

1. Glasgow
2. Newcastle
3. Manchester
4. York
5. Leicester
6. Birmingham
7. Cardiff
8. Exeter
9. Southampton
10. Plymouth

A binary search is to be performed on the names in the list above to locate the name Newcastle.

a Explain why a binary search cannot be performed with the list in its present form.

b Using an appropriate algorithm, alter the list so that a binary search can be performed. State the name of the algorithm you use.

c Use the binary search algorithm on your new list to locate the name Newcastle. *E*

16 *

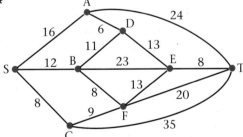

The weighted network shown above models the area in which Bill lives. Each vertex represents a town. The edges represent the roads between the towns. The weights are the lengths, in km, of the roads.

a Use Dijkstra's algorithm to find the shortest route from Bill's home at S to T. Complete all the boxes on the worksheet and explain clearly how you determined the path of least weight from your labelling.

Bill decides that on the way to T he must visit a shop in town E.

b Obtain his shortest route now, giving its length and explaining your method clearly. *E*

17 *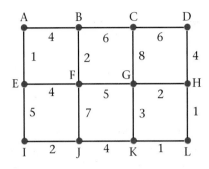

The diagram shows a network of roads. Erica wishes to travel from A to L as quickly as possible. The number on each edge gives the time, in minutes, to travel along that road.

a Use Dijkstra's algorithm to find the quickest route from A to L. Complete all the boxes on the answer sheet and explain clearly how you determined the quickest route from your labelling.

b Show that there is another route which also takes the minimum time. *E*

18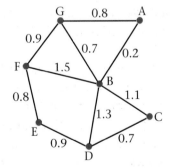

An engineer needs to check the state of a number of roads to see whether they need resurfacing. The roads that need to be checked are represented by the arcs in the diagram. The number on each arc represents the length of that road in km. To check all the roads, he needs to travel along each road at least once. He wishes to minimise the total distance travelled.

The engineer's office is at G, so he starts and ends his journey at G.

a Use an appropriate algorithm to find a route for the engineer to follow. State your route and its length.

The engineer lives at D. He believes he can reduce the distance travelled by starting from home and inspecting all the roads on the way to his office at G.

b State whether the engineer is correct in his belief. If so, calculate how much shorter his new route is. If not, explain why not. (**E**)

19 *

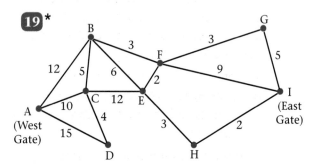

The diagram shows the network of paths in a country park. The number on each path gives its length in km. The vertices A and I represent the two gates in the park and the vertices B, C, D, E, F, G and H represent places of interest.

a Use Dijkstra's algorithm to find the shortest route from A to I. Show all necessary working in the boxes on the worksheet and state your shortest route and its length.

The park warden wishes to inspect each of the paths to check for frost damage. She has to cycle along each path at least once, starting and finishing at *A*.

b i Use an appropriate algorithm to find which paths will be covered twice and state these paths.

ii Find a route of minimum length.

iii Find the total length of this shortest route. (**E**)

20   90 50 55 40 20 35 30 25 45

a Use the bubble sort algorithm to sort the list of numbers above into descending order showing the rearranged order after each pass.

Jessica wants to record a number of television programmes onto video tapes. Each tape is 2 hours long. The lengths, in minutes, of the programmes she wishes to record are:

55 45 20 30 30 40 20 90 25 50 35 and 35

b Find the total length of programmes to be recorded and hence determine a lower bound for the number of tapes required.

c Use the first fit decreasing algorithm to fit the programmes onto her 2-hour tapes.

Jessica's friend Amy says she can fit all the programmes onto 4 tapes.

d Show how this is possible. (**E**)

21

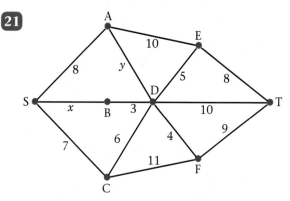

A weighted network is shown above.

Given that the shortest path from S to T is 17 and that $x \geqslant 0$, $y \geqslant 0$:

a i explain why A and C cannot lie on the shortest path,

ii find the value of x.

b Given that $x = 12$ and $y \geqslant 0$, find the possible range of values for the length of the shortest path.

c Give an example of a practical problem that could be solved by drawing a network and finding the shortest path through it. (**E**)

22 * **a** Define the terms
    i tree,
    ii spanning tree,
    iii minimum spanning tree.

b State one difference between Kruskal's algorithm and Prim's algorithm, to find a minimum spanning tree.

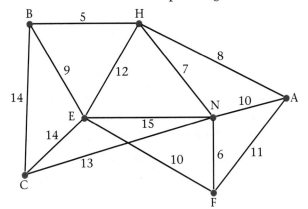

c Use Kruskal's algorithm to find the minimum spanning tree for the network shown above. State the order in which you included the arcs. Draw the minimum spanning tree and state its length.

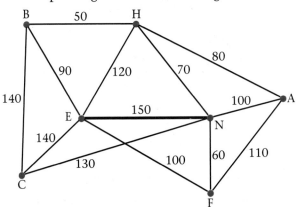

This network models a car park. Currently there are two pay-stations, one at E and one at N. These two are linked by a cable as shown. New pay-stations are to be installed at B, H, A, F and C. The number on each arc represents the distance between the pay-stations in metres. All of the pay-stations need to be connected by cables, either directly or indirectly. The current cable between E and N must be included in the final network. The minimum amount of new cable is to be used.

d Using your answer to part **c**, or otherwise, determine the minimum amount of new cable needed. Draw a diagram to show where these cables should be installed. State the minimum amount of new cable needed.   **E**

23 Nine pieces of wood are required to build a small cabinet. The lengths, in cm, of the pieces of wood are listed below.

    20 20 20 35 40 50 60 70 75

Planks, one metre in length, can be purchased at a cost of £3 each.

a The first fit decreasing algorithm is used to determine how many of these planks are to be purchased to make this cabinet. Find the total cost and the amount of wood wasted.

Planks of wood can also be bought in 1.5 m lengths, at a cost of £4 each. The cabinet can be built using a mixture of 1 m and 1.5 m planks.

b Find the minimum cost of making this cabinet. Justify your answer.   **E**

24 *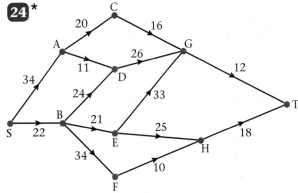

a Use Dijkstra's algorithm to find the shortest route from S to T in this network. Show all necessary working by drawing a diagram. State your shortest route and its length.

b Explain how you determined the shortest route from your labelling.

c It is now necessary to go from S to T via H. Obtain the shortest route and its length.

25 * An algorithm is described by the flow chart below.

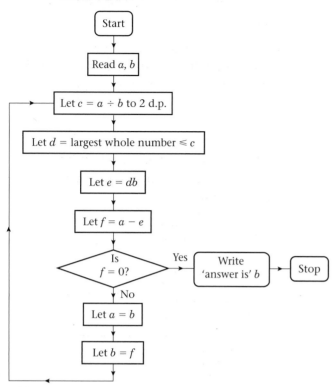

a Given that $a = 645$ and $b = 255$, draw a table to show the results obtained at each step when the algorithm is applied.

b Explain how your solution to part **a** would be different if you had been given that $a = 255$ and $b = 645$.

c State what the algorithm achieves. **E**

26   55 80 25 84 25 34 17 75 3 5

a The list of numbers above is to be sorted into descending order. Perform a bubble sort to obtain the sorted list, giving the state of the list after each complete pass.

The numbers in the list represent weights, in grams, of objects which are to be packed into bins that hold up to 100 g.

b Determine the least number of bins needed.

c Use the first-fit decreasing algorithm to fit the objects into bins which hold up to 100 g. **E**

27 *

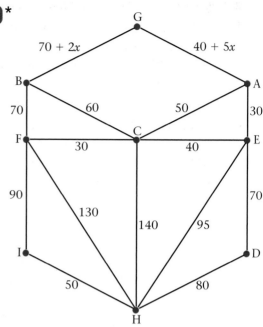

Peter wishes to minimise the time spent driving from his home at H, to a campsite at G. The network above shows a number of towns and the time, in minutes, taken to drive between them. The volume of traffic on the roads into G is variable, and so the length of time taken to drive along these roads is expressed in terms of x, where $x \geq 0$.

a Use Dijkstra's algorithm to find two routes from H to G (one via A and one via B) that minimise the travelling time from H to G. State the length of each route in terms of x.

b Find the range of values of x for which Peter should follow the route via A. **E**

28 *

	A	B	C	D	E	F
A	–	7	3	–	8	11
B	7	–	4	2	–	7
C	3	4	–	5	9	–
D	–	2	5	–	6	3
E	8	–	9	6	–	–
F	11	7	–	3	–	–

The matrix represents a network of roads between six villages A, B, C, D, E and F. The value in each cell represents the distance, in km, along these roads.

a Show this information on a diagram.

b Use Kruskal's algorithm to determine the minimum spanning tree. State the order in which you include the arcs and the length of the minimum, spanning tree. Draw the minimum spanning tree.

c Starting at D, use Prim's algorithm on the matrix given to find the minimum spanning tree. State the order in which you include the arcs. **E**

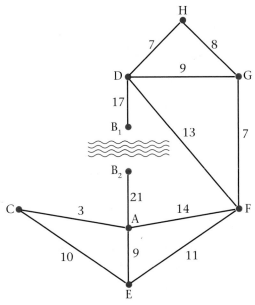

The diagram shows a network of roads connecting villages. The length of each road, in km, is shown. Village B has only a small footbridge over the river which runs through the village. It can be accessed by two roads, from A and D.

The driver of a snowplough, based at F, is planning a route to enable her to clear all the roads of snow. The route should be of minimum length. Each road can be cleared by driving along it once. The snowplough cannot cross the footbridge.

Showing all your working and using an appropriate algorithm,

a find the route the driver should follow, starting and ending at F, to clear all the roads of snow. Give the length of this route.

The local authority decides to build a road bridge over the river at B. The snowplough will be able to cross the road bridge.

b Reapply the algorithm to find the minimum distance the snowplough will have to travel (ignore the length of the new bridge). **E**

30* This diagram describes an algorithm in the form of a flow chart, where a is a positive integer.

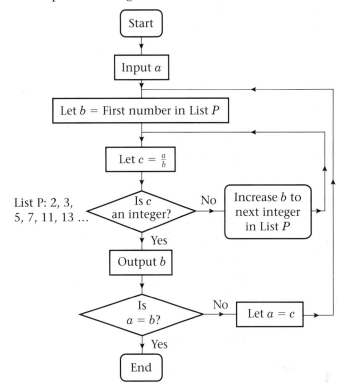

List P, which is referred to in the flow chart, comprises the prime numbers 2, 3, 5, 7, 11, 13, 17, ...

a Starting with $a = 90$, implement this algorithm. Show your working in a table.

b Explain the significance of the output list.

c Write down the final value of c for any initial value of a. **E**

31 * The network opposite represents the journey time, in minutes, between ten Midland towns.

a Use Dijkstra's algorithm to find the quickest route between A and J. Your solution must indicate clearly how you applied the algorithm, including

   i the order in which the vertices were labelled,

   ii how you determined your quickest route from your labelling.

b Is the route you have found the only quickest route? Give a reason for your answer. **E**

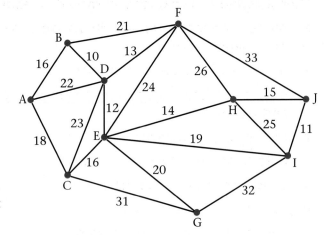

After studying this chapter you should be able to:

1. model a project by an activity network from a precedence table
2. understand the use of dummies
3. carry out a forward pass and a backward pass using early and late event times
4. identify critical activities
5. determine the total float of activities
6. construct cascade (Gantt) charts
7. use cascade (Gantt) charts
8. construct a scheduling diagram.

Critical path analysis

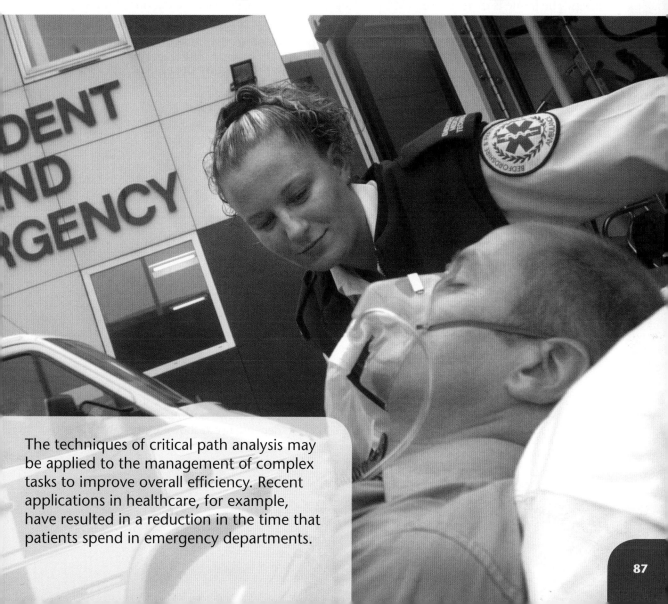

The techniques of critical path analysis may be applied to the management of complex tasks to improve overall efficiency. Recent applications in healthcare, for example, have resulted in a reduction in the time that patients spend in emergency departments.

5.1 You can model a project by an activity network, from a precedence table.

■ Imagine that you have responsibility for the completion of a complex project. The total amount of work to be done is divided into separate activities and some of these cannot be started until others have been completed. You are expected to organise the activities efficiently in order to avoid any unnecessary delay.

■ In order to plan the project effectively you would need to represent the activities in some way that makes any dependencies clear. It would be helpful to make use of some suitable notation and to apply a systematic approach.

■ The activities may be written into a table which shows which ones must be completed before others are started. The table is called a **precedence table**, or, sometimes, a **dependence table**.

Example 1

The manufacture of a sofa involves the construction of a wooden frame which is then fitted with springs. The frame is then covered with padding and material. Cushions are cut out of the same material which must then be stitched and filled. The assembly of the sofa is completed by attaching the cushions. It is then inspected before being wrapped in a protective covering ready for shipping.

Represent the information above in a systematic way that makes any dependencies clear.

This process may be broken down into separate activities.
Typically, these are labelled A, B, C, D, ... for ease of reference later.

A   Build wooden frame
B   Cut out material for cushions
C   Stitch and fill cushions
D   Attach springs to wooden frame
E   Cover frame
F   Complete assembly
G   Inspect
H   Wrap

The activities may now be written in a precedence table.

Activity	Depends on
A	—
B	—
C	B
D	A
E	D
F	C, D
G	F
H	G

E depends on D. The fact that D depends on A is clear from the entry above and is not shown again i.e. only immediate dependence is shown in the table.

Both activities C and D must be completed before work on activity F may be started.

Example 2

The production of a new textbook may be broken down into activities A to G.

Activities A and B do not depend on any other activities.

Both activities C and D can only be started once A has been completed.

Activity E cannot be started until activity B has been completed and activity F cannot be started until activities C and E have been completed. Activity G can only begin once all other activities have been completed.

Draw a precedence table to represent this information.

Activity	Depends on
A	—
B	—
C	A
D	A
E	B
F	C, E
G	D, F

This appears to be the tricky part to complete. It's a lot easier when you realise that you must include just those activities that are not already written in this column.

■ The production of a precedence table goes some way towards representing a project in a form that can help you to coordinate the activities effectively. However, a diagram may be a lot easier to understand, especially if the project is more complex.

■ You have already seen (in Chapter 3) how network diagrams may be used to represent and help analyse a variety of problem types. The information provided in a precedence table may be transferred to an activity network to give a visual representation of the project.

■ There are two types of activity network but only the **activity on arc** type will be used here.

In an activity on arc network, the activities are represented by arcs and the completion of those activities, known as events, are shown as nodes.

- Each arc is labelled with an activity letter.
  The beginning and end of an activity are shown at the ends of the arc and an arrow is used to define the direction.
  The convention is to use straight lines for arcs.

- The nodes are numbered starting from 0 for the first node which is called the **source node**.

- Number each node as it is added to the network.

- The final node is called the **sink node**.

Example 3

Draw an activity network for the precedence table given in Example 2.

Activities A and B do not depend on any previous activities.

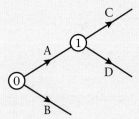

It often isn't easy to get the layout of the network right first time. It's a good idea to draw the network in pencil and have an eraser to hand!

Activities C and D both depend on activity A.
Node 1 represents the end of activity A.

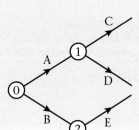

Activity E is dependent on activity B.

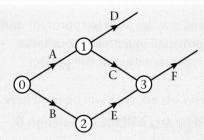

Activity F is dependent on activities C and E. It is now clear that we need to swap the positions of activities C and D in the network to show this.

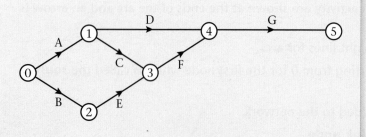

Finally, activity G depends on activities D and F so their arcs need to meet at a node. With practice, it gets easier to anticipate how the network needs to be set out and so there is less need for adjustment.

Exercise 5A

1 The steps involved in starting a car and moving forwards in a straight line are given below.
A Check that car is in neutral.

B Start engine.

C Depress clutch.

D Select first gear.

E Check that it is safe to move off.

F Release the handbrake.

G Raise the clutch and depress the accelerator.
Draw a precedence table for this process.
(There is more than one possible solution.)

2 The development of a commercial computer program is divided into activities A to J.

Activity A does not depend on any other activity.
Activities B, C and D all require that Activity A is completed before they can start.
Activities E and F depend on activity B.
Activity G cannot be started until activities C and E have been completed.
Activity H requires the completion of activity D, while activity I requires that both activities F and G are completed first.
Activity J requires the completion of all activities before it may be started.

a Draw a precedence table to represent the development of the computer program.

b Use the precedence table to draw the corresponding activity network.

3 The precedence table for a project is shown below.
Draw the corresponding activity network.

Activity	Depends on
A	—
B	—
C	A
D	A
E	B
F	B
G	D
H	D
I	C, E
J	F
K	G, I, J
L	H, K

4 Here is an activity network for a project.

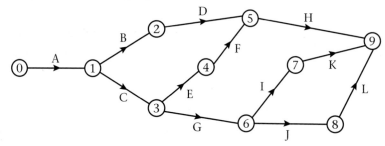

Draw a precedence table to represent the project.

5.2 You need to understand the use of dummies.

■ The precedence table given below appears to be very simple and yet the corresponding activity network cannot be completed using the methods described so far.

Activity	Depends on
A	—
B	—
C	A, B
D	A

Activities A and B do not depend on any other activities and so they are linked to the source node.

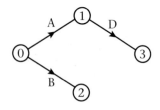

The problem here is how to represent activity C which depends on *both* activity A and activity B.

■ To resolve this problem we introduce what is called a **dummy activity** between events 1 and 2. The dummy activity has no time or cost and its sole purpose, in this case, is to show that activity C depends on activity A as well as activity B.

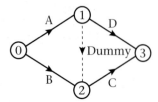

A dummy activity is shown using a dotted line. The direction of the arrow is important. It shows that activity A immediately precedes activity C.

Example 4

Draw an activity network using activity on arc for this precedence table. Use exactly *two* dummies.

Activity	Immediately preceding activities
A	—
B	A
C	A
D	A
E	B
F	B, C
G	D, F
H	D
I	G, H

The label for this column may be written in different ways.

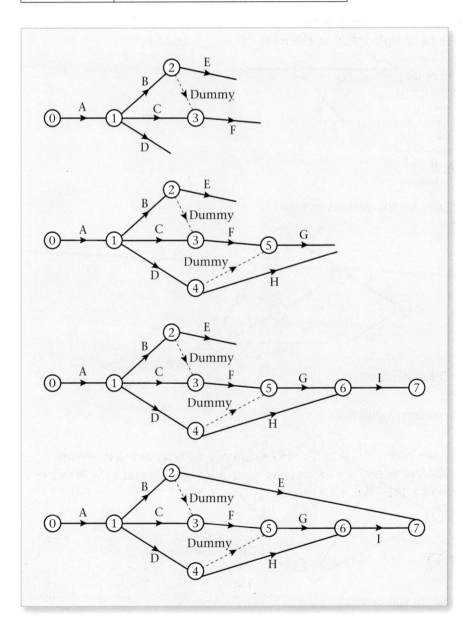

Activity E depends on activity B only, but activity F depends on *both* activity B and activity C. This indicates the need for a dummy.

The first dummy is shown between nodes 2 and 3.

Activity H depends on activity D only, but activity G depends on *both* activity D and activity F. This indicates the need for a second dummy.

Activity I depends on activities G and H.

Extend the arc for activity E to the sink node so that there is only one end point.

■ Every activity must be *uniquely represented* in terms of its events. This requires that there can be *at most one activity* between any two events. Once again, a dummy may be required to satisfy this condition. Here is an example to show how this works.

This diagram shows part of an activity network. Suppose that event S depends on activities P, Q and R.

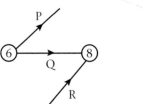

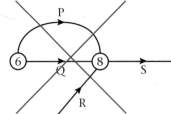

This is not allowed because there are two activities between events 6 and 8.

Using a dummy allows the dependence to be shown while ensuring that all activities are uniquely determined by their events.

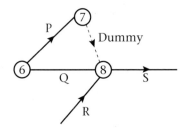

Exercise 5B

1 Draw the precedence table for this activity network.

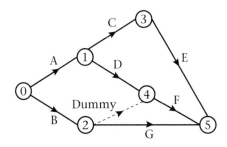

Explain the purpose of the dummy.

2 This activity network contains a dummy.

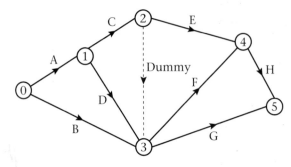

Draw a precedence table for the network.

3 Draw an activity on arc network to represent the precedence table below.

Your network should contain exactly one dummy.

Activity	Must be preceded by
A	–
B	–
C	–
D	A
E	C
F	A, B, E
G	C
H	D

4 Draw an activity on arc network to represent the precedence table below.

Your network should contain exactly two dummies.

Activity	Depends on
A	—
B	—
C	A, B
D	B
E	B
F	C
G	C, D
H	E

5 Draw an activity on arc network for this precedence table using exactly two dummies. Explain the purpose of each dummy.

Activity	Depends on
P	—
Q	—
R	P
S	P
T	P, Q

5.3 You need to be able to carry out a forward pass and a backward pass using early and late event times.

■ So far, activity networks have been used to represent the information given in a precedence table.

■ Each activity takes a certain amount of time to complete, which is referred to as the **duration of the activity**. The next step is to adapt the activity network to allow us to work with this information.

The activity network opposite was used in Exercise 5B Question 1 but now each activity has a figure in brackets representing its duration in hours.

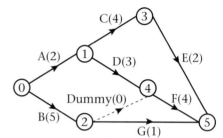

> Activity A takes 2 hours to complete.
> The dummy takes 0 hours to complete.

■ Each node (vertex), of the network represents an event. It is useful to consider two separate times associated with each event.

■ The **early event time** is the earliest time of arrival at the event allowing for the completion of all preceding activities.

■ The **late event time** is the latest time that the event can be left without extending the time needed for the project.

■ The activity network is now adapted to show this information by using ⊟ at each vertex.

> ⊢— Early event time.
> ⊢— Late event time.

■ The early event times are calculated starting from 0 at the source node and working towards the sink node. This is called a **forward pass**, or **forward scan**.

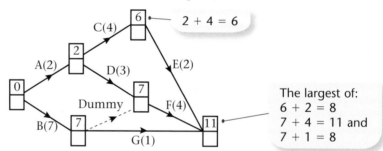

> 2 + 4 = 6

> The largest of:
> 6 + 2 = 8
> 7 + 4 = 11 and
> 7 + 1 = 8

■ The late event times are calculated starting from the sink node and working backwards towards the source node. This is called a **backward pass** or **backward scan**.

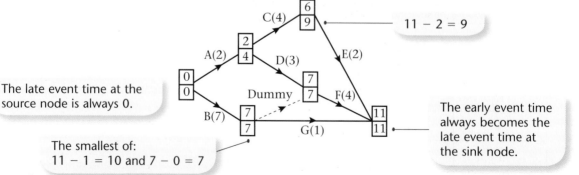

> 11 − 2 = 9

> The late event time at the source node is always 0.

> The early event time always becomes the late event time at the sink node.

> The smallest of:
> 11 − 1 = 10 and 7 − 0 = 7

Example 5

The diagram shows part of an activity network.
Calculate the value of x.

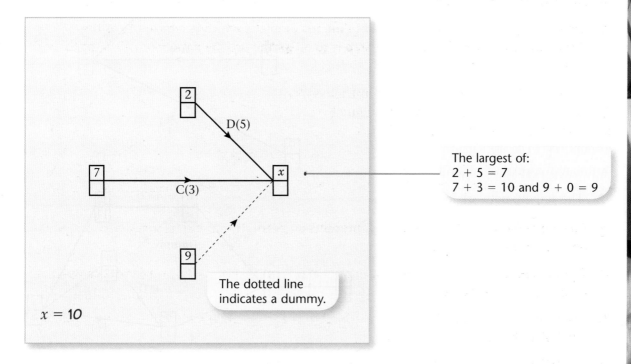

The largest of:
$2 + 5 = 7$
$7 + 3 = 10$ and $9 + 0 = 9$

The dotted line
indicates a dummy.

$x = 10$

Example 6

The diagram shows part of an activity network.
Calculate the value of y.

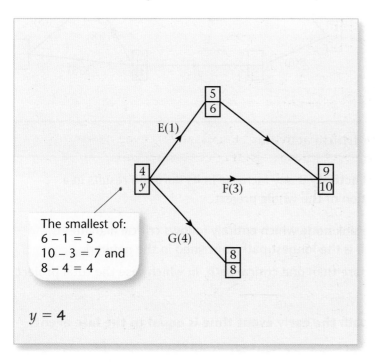

The smallest of:
$6 - 1 = 5$
$10 - 3 = 7$ and
$8 - 4 = 4$

$y = 4$

Exercise 5C

1 The diagram shows part of an activity network. Calculate the value of x.

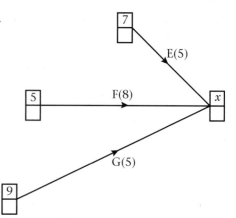

2 The activity network for a project is given opposite.

The time in hours needed to complete each activity is shown in brackets.

Early and late times are shown at each vertex.

Calculate the values of w, x, y and z.

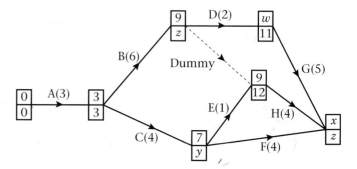

3 The activity network for a project is given opposite.

The time in days needed to complete each activity is shown in brackets.

Calculate the early and late times at each vertex.

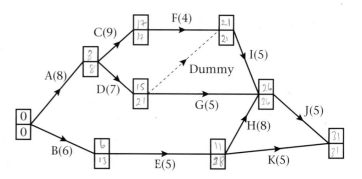

5.4 You need to be able to identify critical activities.

■ An activity is described as a **critical activity** if any increase in its duration results in a corresponding increase in the duration of the whole project.

■ A path from the source node to the sink node which entirely follows critical activities is called a **critical path**. A critical path is the longest path contained in the network.

It is possible for a project to have more than one critical path, in which case the total project time is the same on each one.

■ At each node (vertex) on a critical path **the early event time is equal to the late event time**.

Example 7

The diagram shows an activity network with early and late event times shown at the nodes. Identify the critical activities.

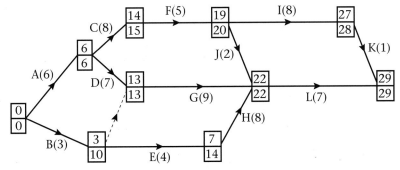

The critical activities are A, D, G, and L. Adding the durations of the critical activities gives the total duration for the project, as shown by the early and late event times at the sink node.

Check: 6 + 7 + 9 + 7 = 29.

Notice that at each vertex on the critical path, the early and late event times are equal. These indicate the critical events. Elsewhere, there is a difference between the two values.

■ **An activity connecting two critical events isn't necessarily a critical activity.**

Example 8

Part of an activity network is shown below including the early and late event times given in hours. Which are the critical activities?

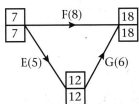

7 + 5 = 12 (critical)
12 + 6 = 18 (critical)
7 + 8 ≠ 18 (not critical)

The critical activities are E and G. F is not a critical activity even though it connects two critical events. Any increase in the duration of activity E or activity G will increase the total time for the project, whereas the duration of activity F may be increased by up to 3 hours and have no effect on the total time.

In the examination, you may need to find more than one critical path and to identify all of the critical activities.

Example 9

Find the critical paths in this activity network and identify the critical activities.

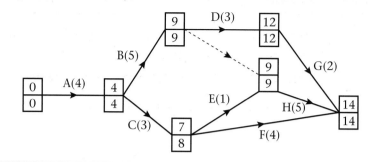

There are *two* critical paths: A B D G and A B H.

The critical activities are A, B, D, G and H.

Exercise 5D

1 Part of an activity network is shown opposite including the early and late event times given in hours.

Activities J and K are critical.

Find the values of x, y and z.

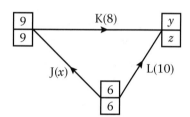

2 The diagram shows an activity network with early and late event times, in hours, shown at the vertices.

a Identify the critical activities.

b Name an activity that links two critical events but is not critical.

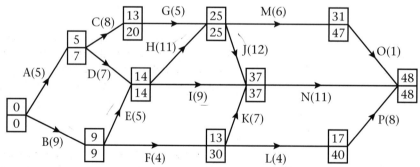

3 The activity network for a project is shown below. Activity times are given in days and are shown in brackets.

a Copy and complete the activity network to show the early and late event times.

b Is G a critical activity?
   Explain your answer.

c Describe the critical path.

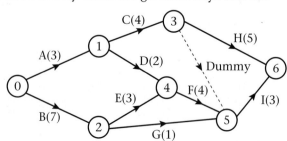

5.5 You can determine the total float of activities.

■ The **total float** of an activity is the amount of time that its start may be delayed without affecting the duration of the project.

total float = latest finish time − duration − earliest start time

■ The total float of any critical activity is 0.

Example 10

Determine the total float of each activity in this activity network.

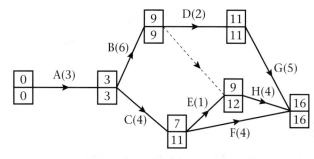

Activity A

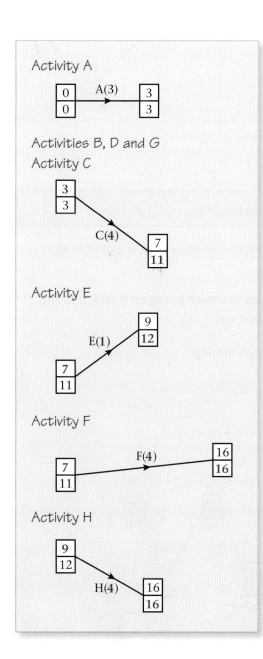

This is a critical activity. Any delay in the start time will affect the duration of the project. Total float = 0.

These are all critical activities so the total float of each one is 0.

The key values here are shown in red.
3 is the earliest time that the activity can start.
11 is the latest time that the activity can be finished by.
4 is the duration of the activity.

total float = 11 − 4 − 3 = 4

7 is the earliest time that the activity can start.
12 is the latest time that the activity can be finished by.
1 is the duration of the activity.

total float = 12 − 1 − 7 = 4

7 is the earliest time that the activity can start.
16 is the latest time that the activity can be finished by.
4 is the duration of the activity.

total float = 16 − 4 − 7 = 5

9 is the earliest time that the activity can start.
16 is the latest time that the activity can be finished by.
4 is the duration of the activity.

total float = 16 − 4 − 9 = 3

Exercise 5E

1 Determine the total float of each activity in this activity network.

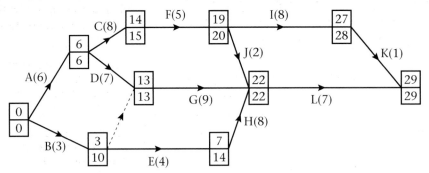

2 The diagram shows part of an activity network with activity times measured in hours.

P is a critical activity.

Q has a total float of 3 hours.

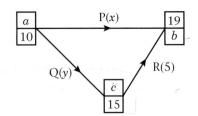

  a Work out the values of a, b, x and y.

  b What is the minimum possible value of c?

  c What is the maximum possible value of the total float of R?

5.6 You need to be able to construct cascade (Gantt) charts.

■ A cascade (Gantt) chart provides a graphical way to represent the range of possible start and finish times for all activities on a single diagram.

■ The number scale shows *elapsed* time. So, the first hour is shown between 0 and 1 on the scale, the second hour is shown between 1 and 2 and so on.

■ The critical activities are shown as rectangles in a line at the top.

Example 11

Here is an activity network for a project.
Early and late event times are shown in hours at the nodes.
Draw a Gantt chart to represent the project.

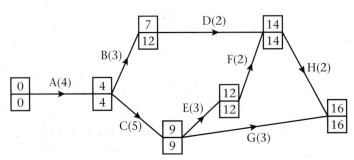

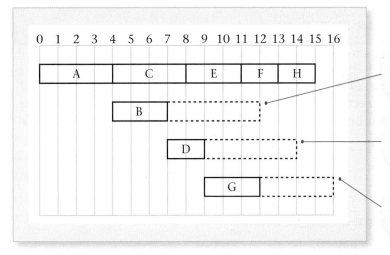

Activity B has duration 3 hours, an earliest start time of 4 hours and a latest finish time of 12 hours.

Activity D has duration 2 hours, an earliest start time of 7 hours and a latest finish time of 14 hours.

Activity G has duration 3, an earliest start time of 9 and a latest finish time of 16.

■ The Gantt chart illustrates clearly that there is no flexibility in the timing of the critical activities. It also illustrates the degree of flexibility in the timing of each of the remaining activities.

■ The total float of each activity is represented by the range of movement of its rectangle on the chart (which is shown as a dotted box).

Exercise 5F

1 The diagram shows an activity network for a project. Early and late event times are shown in days at the nodes. Draw a Gantt chart to represent the project.

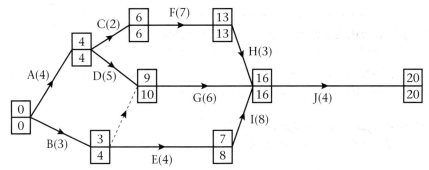

2 An activity network for a project is shown below.

a Calculate the values of w, x, y and z.

b List the critical activities.

c Calculate the total float for activities G and N.

d Draw a Gantt chart to represent the project.

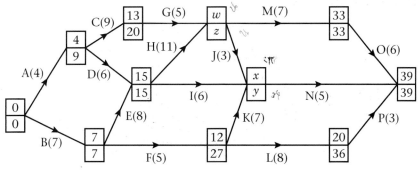

5.7 You need to be able to use cascade (Gantt) charts.

■ The overview of a project provided by a cascade (Gantt) chart allows you to determine which activities *must* be happening at any given time and those that *may* be happening at a given time. In practice, once a project is underway, this provides a useful means of checking that non-critical activities have not been delayed to the point that they have become critical.

Example 12

The Gantt chart below represents a project that must be completed within 25 days.

a Which activities *must* be happening at midday on day 10?

b Which activity *may* be happening on day 10?

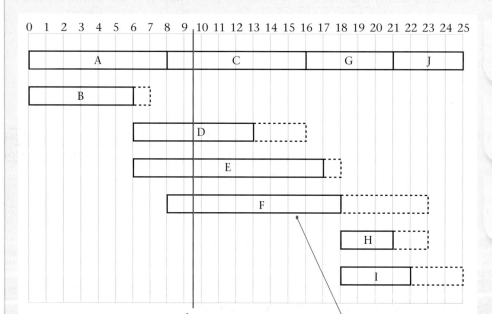

Activity C must be happening at midday on day 10.

Activity D must be happening at midday on day 10.

Activity E must be happening at midday on day 10.

The red line indicates the position of midday on day 10.

Activity F **may** be happening at midday on day 10, but its start could be delayed until day 14 i.e. the end of day 13 on the scale

a The activities that **must** be happening at midday on day 10 are C, D and E.

b The activity that **may** be happening at midday on day 10 is F, because its duration is 10 days and it has to finish at the end of day 23.

Exercise 5G

1 Refer to the Gantt chart shown in Example 12 for this question.

a Which activities *must* be happening at midday on day 8?

b Which activities *must* be happening at midday on day 21?

c Which activities *may* be happening at midday on day 22?

2 The Gantt chart below represents an engineering project. An engineer decides to carry out some spot checks on the progress of the project.

   a Which activities *must* be happening at 12 noon on day 8?

   b Which activities *may* be happening at 12 noon on day 15?

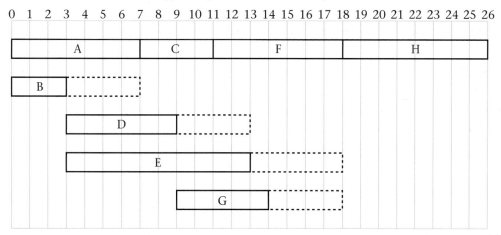

3 Draw a Gantt chart to represent the activity network opposite.

Use your chart to determine:

   a which activities *may* be happening at midday on day 5,

   b which activities *must* be happening at midday on day 7.

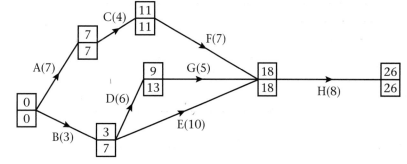

5.8 You need to be able to construct a scheduling diagram.

■ The people responsible for completing the activities in a project are referred to as **workers**. You assume that

   • each activity is completed by a single worker in the time given as the duration of the activity,

   • once an activity has started, it must be completed by the worker,

   • once a worker has completed an activity he/she becomes immediately available to start another activity.

■ You should always use the first available worker.

■ If there is a choice of tasks for a worker, assign the one with the lowest value for its latest finish time as shown on the activity network.

■ The process of assigning workers to activities is known as **scheduling**. Typically, the object is to find the minimum number of workers needed to complete a project in the critical time. Starting from a Gantt chart, a **scheduling diagram** may be constructed which shows the activities assigned to each worker.

Example 13

The diagram shows a Gantt chart for a project.
Schedule the activities to be completed in the critical time by the minimum number of workers.

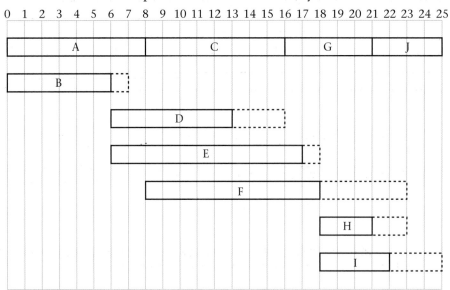

This Gantt chart information may be transferred to a scheduling diagram in stages.

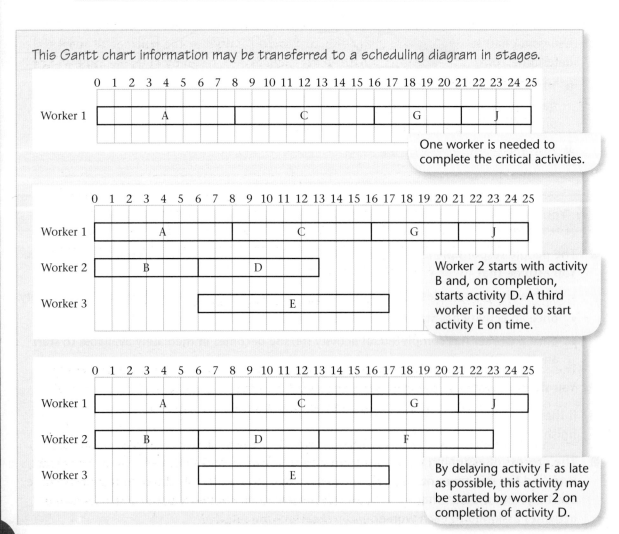

One worker is needed to complete the critical activities.

Worker 2 starts with activity B and, on completion, starts activity D. A third worker is needed to start activity E on time.

By delaying activity F as late as possible, this activity may be started by worker 2 on completion of activity D.

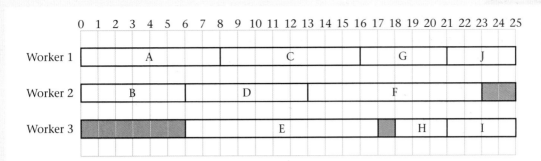

All of the activities have now been assigned and so the scheduling diagram is complete. Shading is used to indicate any periods of inactivity for each worker.

By starting activity H as soon as possible and activity I as late as possible, both activities may be completed by worker 3.

At this stage, if the information is available, it is worth doing a final check to ensure that all of the dependency conditions for the activities are satisfied by the scheduling diagram.

■ In Example 13, the minimum number of workers required to complete the project within the critical time of 25 days was 3. In general, a **lower bound** for the **number of workers needed to complete a project within its critical time** is given by the **smallest integer greater than or equal to**

$$\frac{\text{sum of all of the activity times}}{\text{critical time of the project}}$$

Apply this to Example 13:

Activity	A	B	C	D	E	F	G	H	I	J	Total
Duration	8	6	8	7	11	10	5	3	4	4	66

The sum of all of the activity times is 66 days.

This gives a lower bound as the smallest integer greater than or equal to $\frac{66}{25}$

$\frac{66}{25} = 2.64$ and so a lower bound is 3.

This simply means that it is impossible to complete the project in the critical time using fewer than 3 workers. However, since the calculation takes no account of the degree of overlap of the activities, it doesn't guarantee that 3 workers is sufficient.

■ When the number of available workers is fewer than the minimum required to complete a project within its critical time, the information shown on a Gantt chart cannot be relied upon as some activities will be delayed further than shown. In this situation it is better to construct the scheduling diagram direct from the activity network because special care is needed to ensure that the dependency conditions are satisfied.

Example 14

The diagram shows an activity
network representing a project
with a minimum time of 31 days.
Use a scheduling diagram to find
the new completion time for
the project given that only
two workers are available.

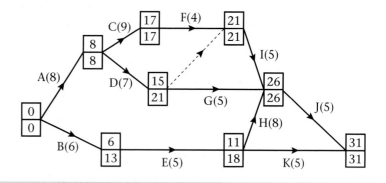

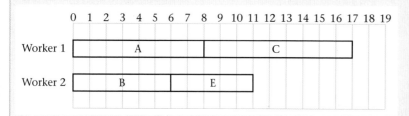

Worker 1 starts activity A and
worker 2 starts activity B.

Worker 2 finishes first and
starts activity E. Worker 1
starts activity C (17 < 21).

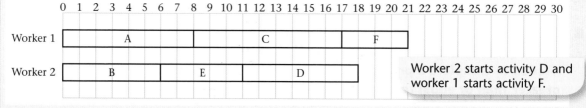

Worker 2 starts activity D and
worker 1 starts activity F.

Continuing in the same way produces the following schedule diagram.

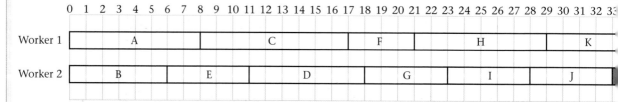

The minimum time to complete the project using two workers is 34 days.

Exercise 5H

1 The cascade chart on the following page represents a project with a critical time of 22 hours.

  a Given that the total duration of all of the activities is 64 hours, calculate a lower bound
for the number of workers needed to complete the project in the minimum time.

  b An unforeseen problem means that Activity B cannot be started until 2 hours into the
project. Does this mean that the time for the whole project is delayed?

  c Which activity *must* be happening 17 hours into the project?

  d Complete a scheduling diagram to complete the project in 22 hours. State the number of
workers required.

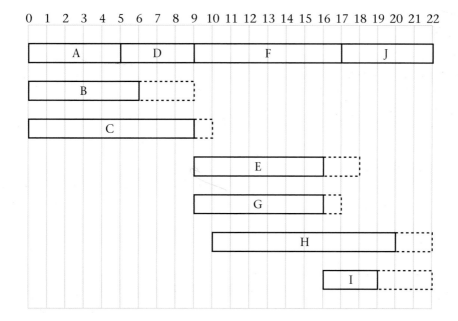

2 The activity network used in Example 14 is shown again here.

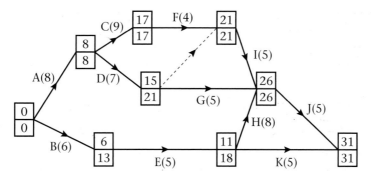

a Draw a Gantt chart to represent the project.

b Schedule the project to be completed by the minimum number of workers in the critical time.

State the number of workers required.

3 Construct a scheduling diagram based on the activity network below, given that only two workers are available.

Find the new minimum time for completion of the project.

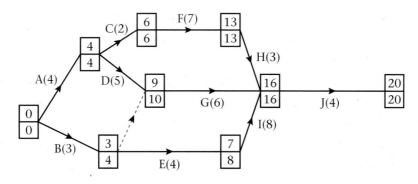

Mixed exercise 5I

1 The precedence table for activities involved in producing a computer game is shown opposite.

An activity on arc network is to be drawn to model this production process.

a Explain why it is necessary to use at least two dummies when drawing the activity network.

b Draw the activity network using exactly two dummies.

E

Activity	Must be preceded by
A	—
B	—
C	B
D	A, C
E	A
F	E
G	E
H	G
I	D, F
J	G, I
K	G, I
L	H, K

2 a Draw the activity network described in this precedence table, using activity on arc and exactly two dummies.

Activity	Immediately preceding activities
A	–
B	–
C	A
D	B
E	B, C
F	B, C

b Explain why each of the two dummies is necessary.

E

3

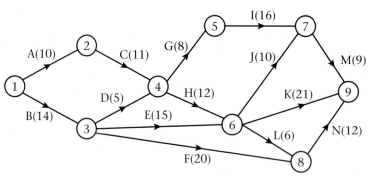

An engineering project is modelled by the activity network shown above. The activities are represented by the arcs. The number in brackets on each arc gives the time, in days, to complete the activity. Each activity requires one worker. The project is to be completed in the shortest time.

a Calculate the early time and late time for each event.

b State the critical activities.

c Find the total float on activities D and F. You must show your working.

d Draw a cascade (Gantt) chart for this project.

The chief engineer visits the project on day 15 and day 25 to check the progress of the work. Given that the project is on schedule,

e which activities *must* be happening on each of these two days?

4

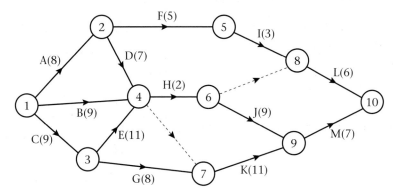

A project is modelled by the activity network shown above. The activities are represented by the arcs. The number in brackets on each arc gives the time, in hours, to complete the activity. The numbers in circles are the event numbers. Each activity requires one worker.

a Explain the purpose of the dotted line from event 6 to event 8.

b Calculate the early time and late time for each event.

c Calculate the total float on activities D, E and F.

d Determine the critical activities.

e Given that the sum of all the times of the activities is 95 hours, calculate a lower bound for the number of workers needed to complete the project in the minimum time. You must show your working.

f Given that workers may not share an activity, schedule the activities so that the process is completed in the shortest time using the minimum number of workers. **E**

5 The network shows the activities that need to be undertaken to complete a project. Each activity is represented by an arc. The number in brackets is the duration of the activity in days. The early and late event times are to be shown at each vertex and some have been completed for you.

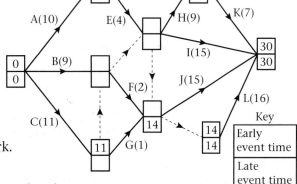

a Calculate the missing early and late times.

b List the two critical paths for this network.

c Explain what is meant by a critical path.

The sum of all the activity times is 110 days and each activity requires just one worker. The project must be completed in the minimum time.

d Calculate a lower bound for the number of workers needed to complete the project in the minimum time. You must show your working.

e List the activities that must be happening on day 20.

f Comment on your answer to part **e** with regard to the lower bound you found in part **d**.

g Schedule the activities, using the minimum number of workers, so that the project is completed in 30 days. **E**

Summary of key points

1   A complex project may be broken down into a number of separate **activities**.
    The completion of an activity is described as an **event**.

2   Some activities may only be started once other activities are completed. This information
    may be shown in a **precedence table**, sometimes called a **dependence table**.

3   An **activity network** provides a visual representation of the information given in a
    precedence table which makes it easier to work with.

    The type of activity network used on this course is called an **activity on arc network**.
    In this type of network, the activities are represented by arcs and events are represented
    as nodes.

4   Each event has an **early event time** and a **late event time**. The early event time is
    the earliest time at which all of the dependent events may be completed. The late event
    time is the latest time at which any of the dependent events may be completed without
    delaying the project. Early event times are calculated using a **forward scan** and late
    event times are calculated using a **backward scan**.

5   An activity is described as a **critical activity** if any increase in its duration results in a
    corresponding increase in the duration of the whole project.

6   A path from the **source node** to the **sink node** which entirely follows critical activities
    is called a **critical path**. A critical path is the longest path contained in the network.

    It is possible for a project to have more than one critical path, in which case the total
    project time is the same on each one.

7   At each vertex on a critical path the early event time is equal to the late event time.

8   The **total float** of an activity is the amount of time that its start may be delayed
    without affecting the duration of the project.

    total float = latest finish time − duration − earliest start time.

    The total float of any critical activity is zero.

9   A **cascade (Gantt) chart** provides a graphical way to represent the range of possible
    start and finish times for all activities on a single diagram.

10 A **scheduling diagram** is used to show how tasks are allocated to workers in order to
    complete a project subject to constraints on the time required or the number of workers
    available.

11 In general, a lower bound for the number of workers needed to complete a project
    within its critical time is given by the smallest integer greater than or equal to

    $$\frac{\text{sum of all of the activity times}}{\text{critical time of the project}}$$

12 A **dummy** activity is shown by a dotted line. It has an arrow to indicate direction, but
    no weight.

13 Dummies are needed for two reasons:
    - for example, if activity D depends *only* on activity B but activity E depends on activies
      B *and* C
    - to enable the unique representation of activities in terms of their end events.

After studying this chapter you should be able to:

1 formulate a problem as a linear programming problem

2 illustrate a two-variable linear programming problem graphically

3 locate the optimal point in a feasible region using the objective line (ruler) method

4 use the vertex testing method to locate the optimal point

5 determine solutions that need integer values.

Linear programming

Linear programming is used to solve problems in a wide variety of areas including finance, planning, control, design, management and resourcing. It is often used to help maximise profits or minimise costs.

This chapter will show you how to formulate a practical problem as a linear programming problem and how to solve a two-variable linear programming problem graphically.

Linear programming can be used to determine how many of each type of floral display to make in order to create the greatest profit.

6.1 You need to be able to formulate a problem as a linear programming problem.

■ The **decision variables** in a linear programming problem are the numbers of each of the things that can be varied. For example, the number of fruit cakes made, the number of teddy bears made, etc. The variables, which are often called x, y, z, etc., will be the 'letters' in the inequalities and **objective function**.

■ The objective function is the aim of the problem. It may be to maximise profit or to minimise cost. There are two parts: a word 'maximise' or 'minimise', and an algebraic expression, which is often written as an equation in terms of the decision variables, for example, $P = 3x + 2y$, $C = 4x + 7y + 3z$.

■ The **constraints** are the things that will prevent you making, or using, an infinite number of each of the variables. Examples of constraints are the quantity of raw materials available, the time available, the fact that you cannot have a negative quantity, etc. **Each constraint will give rise to one inequality**.

■ If you find values for the decision variables that satisfy each constraint you have a **feasible solution**.

■ In a graphical linear programming problem, the region that contains *all* the feasible solutions is called the **feasible region**.

■ The **optimal solution** is the feasible solution that meets the objective. **There might be more than one optimal solution**.

■ To formulate a problem as a linear programming problem:
  1 define the decision variables (x, y, z, etc.)
  2 state the objective (maximise or minimise, together with an algebraic expression)
  3 write the constraints as inequalities.

Example 1

Mrs Cook is making cakes to sell for charity. She makes two types of cake, fruit and chocolate. Amongst other ingredients, each fruit cake requires 1 egg, 250 g of flour and 200 g of sugar. Each chocolate cake requires 2 eggs, 250 g of flour and 300 g of sugar.

Mrs Cook has 36 eggs, 7 kg of flour and 6 kg of sugar.

She will sell the fruit cakes for £3.50 and the chocolate cakes for £5.

She wishes to maximise the money she makes from these sales.

You may assume she sells all she makes.

Formulate this as a linear programming problem.

This is an assumption that is made in linear programming problems and is not usually stated.

It is sometimes useful to summarise the information in a table.

Type of cake	Eggs	Flour	Sugar	Price
Fruit	1	250 g	200 g	£3.50
Chocolate	2	250 g	300 g	£5.00
Total available	36	7000 g	6000 g	

You need to make the units agree.

First define the decision variables.

Let f be the number of fruit cakes made.

Let c be the number of chocolate cakes made.

Next, state the objective.

> The decision variables always start 'Let x be the number of ...' etc.

> Maximise because Mrs Cook wishes to maximise her income.

> P is just a useful way of referring to the expression. Mrs Cook wants to maximise her profit, P.

> There are two parts to the objective: a word maximise or minimise and an algebraic expression, usually in terms of x, y, etc.

$$\text{maximise } P = 3.5f + 5c$$

> Mrs Cook will get £3.50 for each fruit cake she sells. If she sells f of them she will make £3.5f.

> Each chocolate cake sold raises £5. If c cakes are sold, Mrs Cook will make £5c.

> *Note* that there are *no* units, in this case £ s, shown in the objective function.

Finally, identify the constraints

> These will be inequalities.

eggs: $f + 2c \leqslant 36$

> Mrs Cook needs 1 egg for each fruit cake and 2 to make each chocolate cake. There is a maximum of 36 eggs that can be used.

flour: $250f + 250c \leqslant 7000$

This simplifies to

$$f + c \leqslant 28$$

> Each fruit cake requires 250 g and each chocolate cake 250 g. There are up to 7000 g available.

sugar: $200f + 300c \leqslant 6000$

This simplifies to

$$2f + 3c \leqslant 60$$

> Each fruit cake requires 200 g and each chocolate cake 300 g. There are 6000 g available.

non-negativity: $f \geqslant 0 \; c \geqslant 0$

These are often written together as

$$f, c \geqslant 0$$

A formal summary of the problem is

> You cannot have negative value for f or c. (A negative cake is not possible!)

Let f be the number of fruit cakes made.

Let c be the number of chocolate cakes made.

maximise $\qquad P = 3.5f + 5c$

subject to:

$$f + 2c \leqslant 36$$
$$f + c \leqslant 28$$
$$2f + 3c \leqslant 60$$
$$f, c \geqslant 0$$

> This is the **formal template for presenting a linear programming problem**.

Example 2

A company buys two types of diary to send to its customers, a desk top diary and a pocket diary.

They will need to place a minimum order of 200 desk top and 80 pocket diaries.

They will need at least twice as many pocket diaries as desk top diaries.

They will need a total of at least 400 diaries.

Each desk top diary costs £6 and each pocket diary costs £3.

The company wishes to minimise the cost of buying the diaries.

Formulate this as a linear programming problem.

$$y \geqslant 2x$$

Summarise the information in a table.

Type of diary	minimum order	cost
Desk top	200	£6
Pocket	80	£3

Also require:
- twice as many pocket as desk top
- total of at least 400.

Define the decision variables.

Let x be the number of desk top diaries bought.

Let y be the number of pocket diaries bought.

State the objective.

minimise $\quad c = 6x + 3y$

> Each desk top diary costs £6 and each pocket diary costs £3. The company wants to minimise the cost.

State the constraints.

minimum order: $\qquad x \geqslant 200$

$$y \geqslant 80$$

> The company must order at least 200 desk top diaries and at least 80 pocket diaries.

at least twice as many pocket as desk top:

$$2x \leqslant y$$

> This type of comparative constraint can be tricky. It sometimes helps to see it as two steps: first getting the algebra correct and then getting the inequality correct.

> To get the algebra correct, change the statement to read 'exactly twice as many pocket as desk top'. This tells you which one needs to be doubled. The number of pocket diaries (y) is double the number of desk top (x). So we get $2x = y$

> To get the direction of the inequality you need to see which one of these can be made larger. Here the number of pocket diaries can increase so $2x \leqslant y$.

a total of at least 400 diaries:

$$x + y \geqslant 400$$

> The total number of diaries is simply $x + y$.

non-negativity:

Since we have already stated that $x \geqslant 200$ and $y \geqslant 80$, we do not need to state that $x \geqslant 0$ and $y \geqslant 0$.

So the non-negativity constraint is not needed here.

Here is the summary.

> Let x be the number of desk top diaries bought.
>
> Let y be the number of pocket diaries bought.
>
> minimise $\quad c = 6x + 3y$
>
> subject to $\qquad x \geqslant 200$
>
> $\qquad\qquad\quad y \geqslant 80$
>
> $\qquad\qquad 2x \leqslant y$
>
> $\qquad\quad x + y \geqslant 400$

Example 3

A company produces two types of syrup, A and B. The syrups are a blend of sugar, fruit and juice.

Syrup A contains 30% sugar, 50% fruit and 20% juice.

Syrup B contains 20% sugar, 35% fruit and 45% juice.

Each litre of syrup A costs 50p and each litre of syrup B costs 40p.

There is a maximum daily production of 40 000 litres of syrup A and 45 000 litres of syrup B.

A confectionery manufacturer places an order for 60 000 litres of syrup but requires

- below 25% sugar
- at least 40% fruit
- no more than 35% juice.

The company will blend syrups A and B to meet the confectionery manufacturer's requirements.

The company wishes to minimise its costs.

Letting x be the number of litres of syrup A used, and y be the number of litres of syrup B used, formulate this as a linear programming problem.

Summarise the information.

Syrup	sugar %	fruit %	juice %	maximum amount (ℓ)	cost per litre
A (x)	30	50	20	40 000	50p
B (y)	20	35	45	45 000	40p
Required in final blend	< 25	$\geqslant 40$	$\leqslant 35$	60 000	

Define the variables.

This has already been done in the question.

State the objective.

minimise $c = 0.5x + 0.4y$ •————

> Each litre of syrup A costs 50p and each litre of syrup B costs 40p. The company wants to minimise its costs.

> *Note* the strict inequality. It must be *below* 25%.

State the constraints.

> This is another tricky type of constraint. Be careful!

sugar: $0.3x + 0.2y < 0.25 (x + y)$

> 30% of syrup A is sugar.

> 20% of syrup B is sugar.

> at most ...

> ... 25% of the combined blend must be sugar.

This simplifies to $\quad x < y$ •————

> $0.3x + 0.2y < 0.25x + 0.25y$
> $0.05x < 0.05y$

fruit: $0.5x + 0.35y \geqslant 0.4 (x + y)$

> Each litre of syrup A contains 50% fruit.

> Each litre of syrup B contains 35% fruit.

> at least ...

> ... 40% fruit in the combined blend.

This simplifies to $\quad 2x \geqslant y$ •————

> $0.5x + 0.35y \geqslant 0.4x + 0.4y$
> $0.1x \geqslant 0.05y$

juice: $0.2x + 0.45y \leqslant 0.35 (x + y)$

which simplifies to $\quad 2y \leqslant 3x$ •————

> $0.2x + 0.45y \leqslant 0.35x + 0.35y$
> $0.1y \leqslant 0.15x$
> $10y \leqslant 15x$

amount required:

$$x + y \geqslant 60\,000$$

non-negativity:

$$x, y \geqslant 0$$

Here is the summary

> minimise $c = 0.5x + 0.4y$
> subject to
> $\qquad x < y$
> $\qquad 2x \geqslant y$
> $\qquad 3x \geqslant 2y$
> $\qquad x + y \geqslant 60\,000$
> $\qquad x, y \geqslant 0$

Exercise 6A

1 A chocolate manufacturer is producing two hand-made assortments, gold and silver, to commemorate 50 years in business.

It will take 30 minutes to make all the chocolates for one box of gold assortment and 20 minutes to make the chocolates for one box of silver assortment.

It will take 12 minutes to wrap and pack the chocolates in one box of gold assortment and 15 minutes for one box of silver assortment.

The manufacturer needs to make at least twice as many silver as gold assortments.

The gold assortment will be sold at a profit of 80p, and the silver at a profit of 60p.

There are 300 hours available to make the chocolates and 200 hours to wrap them. The profit is to be maximised.

Letting the number of boxes of gold assortment be x and the number of boxes of silver assortment be y, formulate this as a linear programming problem.

2 A floral display is required for the opening of a new building. The display must be at least 30 m long and is to be made up of two types of planted displays, type A and type B.

Type A is 1 m in length and costs £6

Type B is 1.5 m in length and costs £10

The client wants at least twice as many type A as type B, and at least 6 of type B.

The cost is to be minimised.

Letting x be the number of type A used and y be the number of type B used, formulate this as a linear programming problem.

3 A toy company makes two types of board game, Cludopoly and Trivscrab. As well as the board, each game requires playing pieces and cards.

The company uses two machines, one to produce the pieces and one to produce the cards. Both machines can only be operated for up to ten hours per day.

The first machine takes 5 minutes to produce a set of pieces for Cludopoly and 8 minutes to produce a set of pieces for Trivscrab.

The second machine takes 8 minutes to produce a set of cards for Cludopoly and 4 minutes to produce a set of cards for Trivscrab.

The company knows it will sell at most three times as many games of Cludopoly as Trivscrab.

The profit made on each game of Cludopoly is £1.50 and £2.50 on each game of Trivscrab.

The company wishes to maximise its daily profit.

Let x be the number of games of Cludopoly and y the number of games of Trivscrab.

Formulate this problem as a linear programming problem.

4 A librarian needs to purchase bookcases for a new library. She has a budget of £3000 and 240 m^2 of available floor space. There are two types of bookcase, type 1 and type 2, that she is permitted to buy.

Type 1 costs £150, needs 15 m^2 of floor space and has 40 m of shelving.

Type 2 costs £250, needs 12 m^2 of floor space and has 60 m of shelving.

She must buy at least 8 type 1 bookcases and wants at most $\frac{1}{3}$ of all the bookcases to be type 2.
She wishes to maximise the total amount of shelving.

Letting x and y be the number of type 1 and type 2 bookcases bought respectively, formulate this as a linear programming problem.

5 A garden supplies company produces two different plant feeds, one for indoor plants and one for outdoor plants.

In addition to other ingredients, the plant feeds are made by combining three different natural ingredients A, B and C.

Each kilogram of indoor feed requires 10 g of A, 20 g of B and 20 g of C.

Each kilogram of outdoor feed requires 20 g of A, 10 g of B and 20 g of C.

The company has 5 kg of A, 5 kg of B and 6 kg of C available each week to use to make these feeds.

The company will sell at most three times as much outdoor as indoor feed, and will sell at least 50 kg of indoor feed.

The profit made on each kilogram of indoor and outdoor feed is £7 and £6 respectively. The company wishes to maximise its weekly profit.

Formulate this as a linear programming problem, defining your decision variables.

6 Sam makes three types of fruit smoothies, A, B and C. As well as other ingredients all three smoothies contain oranges, raspberries, kiwi fruit and apples, but in different proportions. Sam has 50 oranges, 1000 raspberries, 100 kiwi fruit and 60 apples. The table below shows the number of these 4 fruits used to make each smoothie and the profit made per smoothie. Sam wishes to maximise the profit.

Smoothie	Oranges	Raspberries	Kiwi fruit	Apples	Profit
A	1	10	2	2	60p
B	$\frac{1}{2}$	40	3	$\frac{1}{2}$	65p
C	2	15	1	2	55p
Total available	50	1000	100	60	

Letting x be the number of A smoothies, y the number of B smoothies and z the number of C smoothies, formulate this as a linear programming problem.

7 A dairy manufacturer has two factories R and S. Each factory can process milk and yoghurt.

Factory R can process 1000 litres of milk and 200 litres of yoghurt per hour.

Factory S can process 800 litres of milk and 300 litres of yoghurt per hour.

It costs £300 per hour to operate factory R and £400 per hour to operate factory S. In order to safeguard jobs it has been agreed that each factory will operate for at least $\frac{1}{3}$ of the total, combined, operating time.

The manufacturer needs to process 20 000 litres of milk and 6000 litres of yoghurt. He wishes to distribute this between the 2 factories in such a way as to minimise operating costs.
Formulate this as a linear programming problem in x and y, defining your decision variables.

6.2 You can illustrate a two-variable linear programming problem graphically.

■ Equations of the form $ax + by = c$ or $y = mx + c$ are called **linear equations**, because their graphs are straight lines.

■ Straight line graphs *must* be drawn with a ruler.

■ Once a straight line graph has been drawn there are three regions.

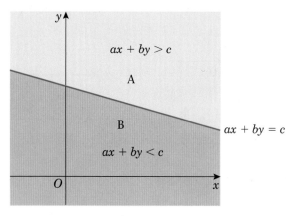

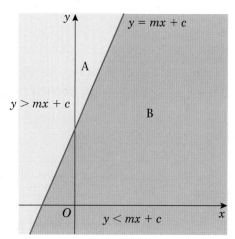

- All the points that lie on the line are represented by
    $$ax + by = c \text{ or } y = mx + c$$
- All the points that lie above the line, in region A, are represented by
    $$ax + by > c \text{ or } y > mx + c$$
- All the points that lie below the line, in region B, are represented by
    $$ax + by < c \text{ or } y < mx + c$$

> If the line is to be included in the region we use $\geq$ or $\leq$ in place of $>$ or $<$. So points region A, *including the line*, would be represented as $ax + by \geq c$ or $y \geq mx + c$.

■ When illustrating an inequality, first draw a straight line and then use shading
- **Strict inequalities** (using $<$ or $>$) are represented by a **dashed** (or **dotted**) **line**. This indicates that the **line itself** is *not* **included in the region**.
- Inequalities using $\leq$ or $\geq$ are represented by a **solid line** to show that the **line is included in the region**.

■ The convention is that the region that does *not* satisfy the inequality is shaded.

> To decide which side of the line does not satisfy the inequality, choose a convenient point in each region and substitute its coordinates into the inequality to see whether (or not) it satisfies the inequality. Shade the region containing the point that does *not* satisfy the inequality.

■ The feasible region satisfying several inequalities will therefore be the one left unshaded.

This is the feasible region for these three inequalities.

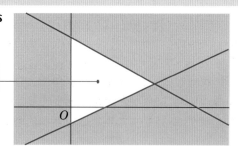

Example 4

Write down the inequalities shown by regions A, B, C, D, E, F, G, H, I and J in the diagrams below.

a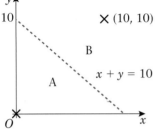

b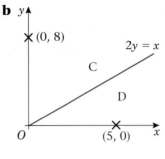

c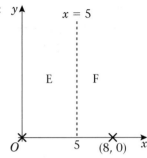

d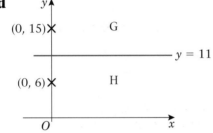

e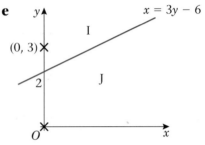

a Region A $x + y < 10$ •———— A broken line, so < and > rather than ⩽ and ⩾.

     └———— Testing point (0, **0**) **0** + **0** < 10.

     Region B $x + y > 10$ •

     └———— Testing point (**10**, **10**) **10** + **10** > 10.

b Region C $2y \geqslant x$ •———— A solid line, so ⩽ and ⩾ rather than < and >.

     └———— Testing (0, **8**) 2 × **8** ⩾ 0.

     Region D $2y \leqslant x$ •

     └———— Testing (**5**, 0) 2 × **0** ⩽ 5.

c Region E $x < 5$ •———— A broken line, so a strict inequality.

     └———— Testing (0, **0**) **0** < 5.

     Region F $x > 5$ •

     └———— Testing (**8**, 0) **8** > 5.

d Region G $y \geqslant 11$ •———— A solid line.

     └———— Testing (0, **15**) gives **15** ⩾ 11.

     Region H $y \leqslant 11$ •

     └———— Testing (0, **6**) gives **6** ⩽ 11.

e Region I $x \leqslant 3y - 6$ •———— A solid line.

     └———— Testing (0, **3**) **0** ⩽ 3 × **3** − 6.

     Region J $x \geqslant 3y - 6$ •

     └———— Testing (0, **0**) **0** ⩾ 3 × **0** − 6.

$x = 3y - 6$ could be written in many different forms, such as $3y = x + 6$ or $3y - x = 6$. You must use a coordinate to check the direction of the inequality since region I is given by

$$x \leqslant 3y - 6$$

or $\qquad 3y \geqslant x + 6$

or $\qquad 3y - x \geqslant 6$

Here the inequality seems to be reversed in the second and third line, but the side including the positive y term remains on the 'greater than' side.

Example 5

Illustrate on a diagram the region, R, for which

$$x \geqslant 2$$
$$4x + 3y < 12$$
$$2y \leqslant x$$
$$x, y \geqslant 0$$

Label the region R.

There are five inequalities here and three lines to add to the axes.

$x \geqslant 2$

Draw $x = 2$.

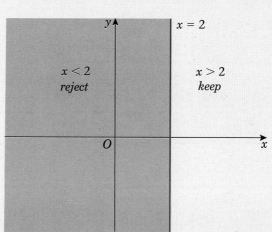

$4x + 3y < 12$

Draw $4x + 3y = 12$ as a broken line.

When $x = 0$, $y = 4$.

When $y = 0$, $x = 3$.

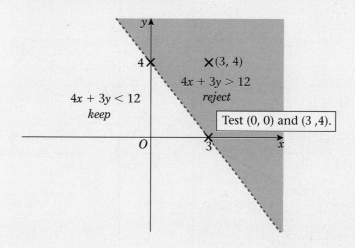

Test $(0, 0)$ and $(3, 4)$.

$2y \le x$

The line $2y = x$ passes through $(0, 0)$, $(2, 1)$, $(4, 2)$, $(6, 3)$.

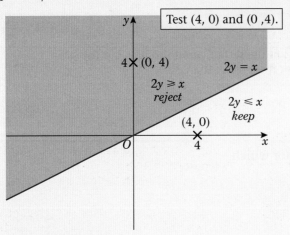

Test $(4, 0)$ and $(0, 4)$.

$4 \times (0, 4)$

$2y \ge x$
reject

$2y = x$

$2y \le x$
keep

$(4, 0)$
4

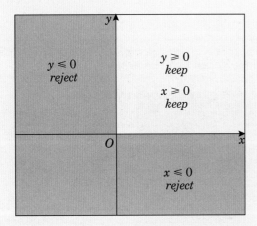

$y \le 0$
reject

$y \ge 0$
keep

$x \ge 0$
keep

$x \le 0$
reject

Combine all these on one diagram.

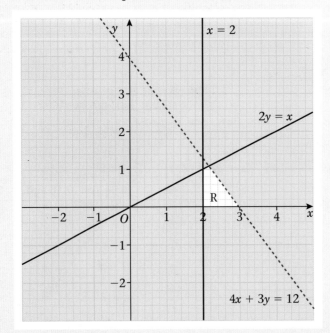

$x = 2$

$2y = x$

R

$4x + 3y = 12$

Colour has been used here for clarity. Do not use colour in the examination!

Example 6

Illustrate the inequalities in Example 3. Indicate the feasible region by labelling it R.

In addition to the non-negativity inequalities, $x, y \geq 0$, the inequalities are:
$x < y$, $2x \geq y$, $3x \geq 2y$, $x + y \geq 60\,000$.

$x < y$

Draw $x = y$ (broken) passing through $(0, 0)$, $(10\,000, 10\,000)$, $(60\,000, 60\,000)$

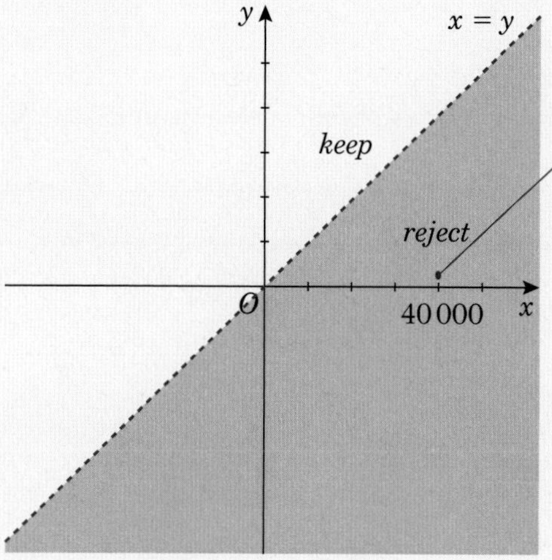

Testing $(40\,000, 0)$ puts it in the region to *reject* since $40\,000 \not< 0$.

$2x \geq y$

The line $2x = y$ passes through $(0, 0)$, $(10\,000, 20\,000)$, $(30\,000, 60\,000)$.

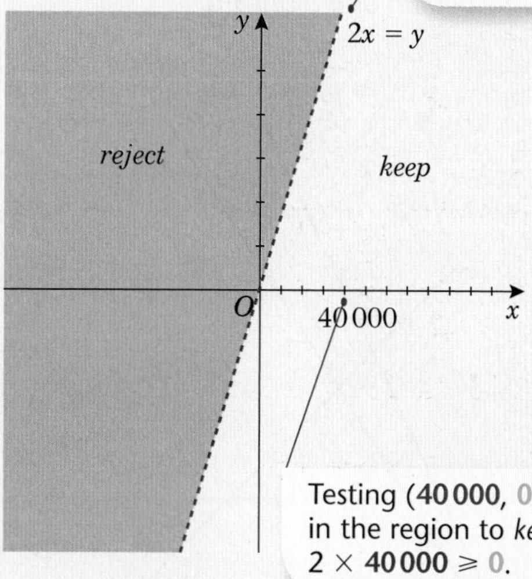

Testing $(40\,000, 0)$ puts it in the region to *keep* since $2 \times 40\,000 \geq 0$.

$3x \geq 2y$

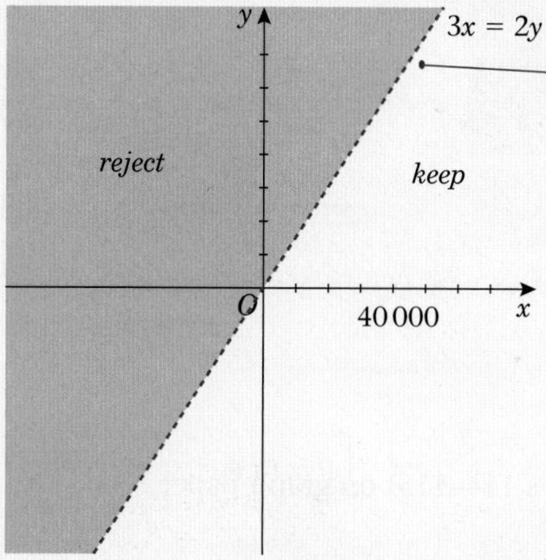

The line $3x = 2y$ passes through $(0, 0)$, $(20\,000, 30\,000)$, $(40\,000, 60\,000)$.

Testing $(40\,000, 0)$ puts it in the region to *keep* since $3 \times 40\,000 \geq 2 \times 0$.

$x + y \geqslant 60\,000$

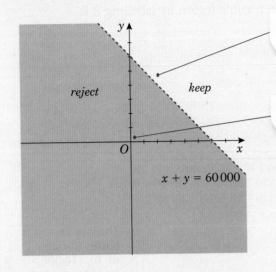

The line $x + y = 60\,000$ passes through $(0, 60\,000)$ and $(60\,000, 0)$.

Testing $(0, 0)$ puts it in the region to *reject* since $0 + 0 \not\geqslant 60\,000$.

Combine all these on one diagram.

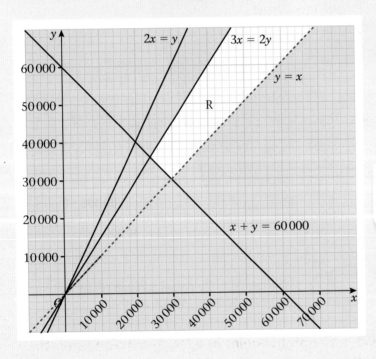

The inequalities $2x \geqslant y$, $x \geqslant 0$ and $y \geqslant 0$ are not actually used in forming the feasible region.

Exercise 6B

1. Illustrate the inequalities found in Example 1 (pages 114–115) on graph paper. Label the feasible region, R.

2. Repeat Question 1 using the inequalities found in Example 2 (pages 116–117).

3 Represent the following inequalities graphically. Label the region bounded by these inequalities R.

$$2x + 3y > 18$$
$$y > x$$
$$y \leqslant 5$$
$$x, y \geqslant 0$$

4 The following inequalities were found when solving a linear programming problem.

$$2x \geqslant 3y$$
$$3x + 4y \leqslant 24$$
$$x \geqslant 3$$
$$y \geqslant 1$$

Represent these inequalities on a graph.

Indicate the feasible region by labelling it R.

5 Region R is bounded by the following inequalities

$$x + y \leqslant 20$$
$$5x + 6y \geqslant 60$$
$$2x \geqslant y$$
$$y \leqslant 10$$

By drawing suitable straight lines, draw a graph to show region R.

6 Indicate on a graph the region, R, for which

$$2x - 3 < y$$
$$y > 3$$
$$y > 6 - 2x$$
$$x \geqslant 0$$

6.3 **You need to be able to locate the optimal point in a feasible region using the objective line (ruler) method.**

- You know how to identify the feasible region using straight line graphs and shading.

- Since all of the points in the feasible region obey the constraints, they are all potential solutions to the linear programming problem.

- You need to be able to identify which solution(s) fulfil(s) the objective.

- There are two methods you need to know to do this.

- The first is called the **objective line method (ruler method)**.

Example 7

Nigel is making ice cream for sale at a charity fair. He makes two flavours of ice cream: vanilla and chocolate. Let the number of litres of vanilla ice cream made be x and the number of litres of chocolate ice cream made be y. Nigel decides to use linear programming to determine the number of litres of each type of ice cream he should make. The constraints and the feasible region, R, are illustrated in the diagram below.

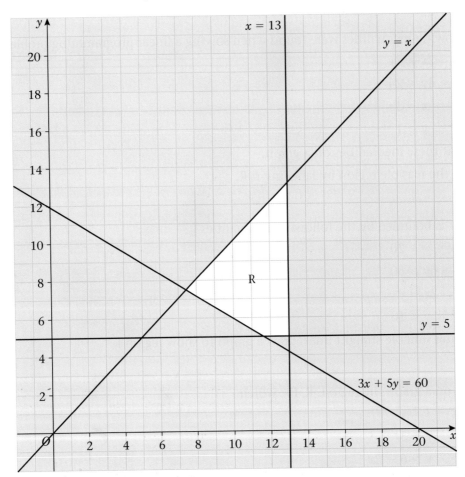

Determine the optimal solution for this problem, given that the objective is to

a maximise the profit on sales, $P = 2x + y$,

b minimise the production costs, $C = 5x + 2y$.

a The diagram shows the parallel lines of the form $P = 2x + y$, for various values of P.
These are all **objective lines**. The value of P is given for each line.
The value of P **increases as the parallel lines move to the right**, away from the origin.

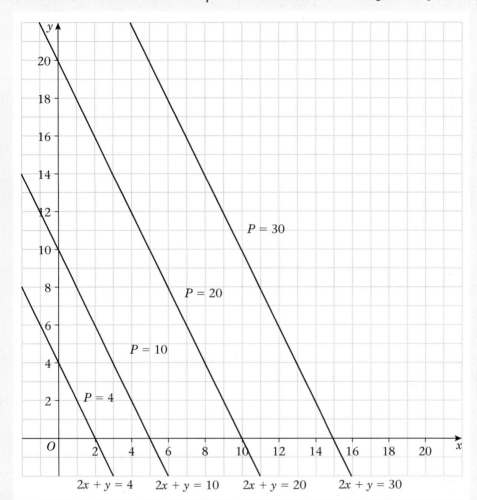

Imagine a ruler sliding over the feasible region on page 128 so that it is always parallel to the profit lines above. The maximum value of P will be in the feasible region at the point furthest from the origin (the last point the ruler touches as it slides out of the feasible region).

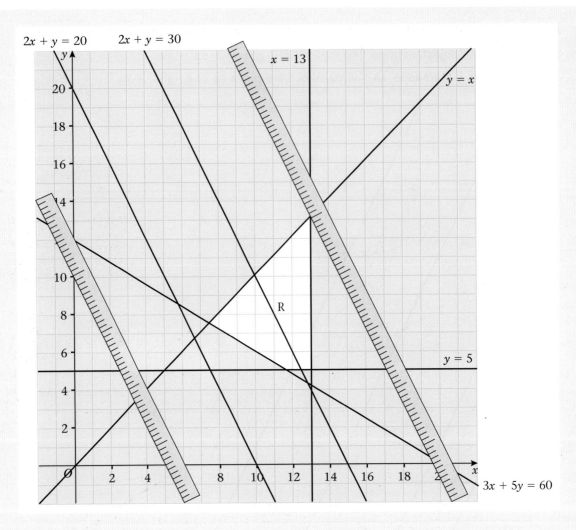

From the diagram, the optimal point is (13, 13), giving an optimal value for P of
2 × 13 + 13 = 39.

So Nigel should make 13 litres of vanilla ice cream,
13 litres of chocolate ice cream, and makes
a profit of £39.

You should always make
sure you give a clear answer
to the question asked.

b Use a ruler to draw an objective line. Then slide the ruler towards the feasible region, keeping it parallel to the objective line.

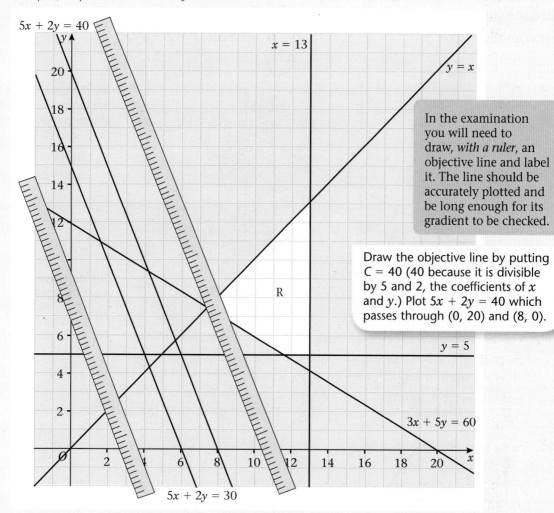

$5x + 2y = 40$

$x = 13$

$y = x$

In the examination you will need to draw, *with a ruler*, an objective line and label it. The line should be accurately plotted and be long enough for its gradient to be checked.

Draw the objective line by putting $C = 40$ (40 because it is divisible by 5 and 2, the coefficients of x and y.) Plot $5x + 2y = 40$ which passes through (0, 20) and (8, 0).

R

$y = 5$

$3x + 5y = 60$

$5x + 2y = 30$

The minimum value will occur at the first point covered by the objective line as it moves into the feasible region. In this case the optimal point is found where the line $y = x$ meets the line $3x + 5y = 60$. Solving these simultaneously $8x = 60$ so $x = 7.5$ and $y = 7.5$. Since $C = 5x + 2y$, $C = 7 \times 7.5 = 52.5$.

So Nigel should make 7.5 litres of vanilla ice cream, 7.5 litres of chocolate ice cream, with production costs of £52.50.

■ For a **maximum point**, look for the **last point covered by an objective line** as it **leaves** the **feasible region**.

■ For a **minimum point**, look for the **first point covered by an objective line** as it **enters** the **feasible region**.

■ It is very important, when using the objective line method, that the ruler is kept parallel to an objective line. To do this you need two straight edges: a ruler and either a second ruler or a set square.

Place the ruler along an objective line, then place the set square (or second ruler) at the base of the ruler.

ruler

set square

Hold the set square firmly and slide the ruler along the edge of the set square.

Example 8

Using the feasible region found in Example 5, find the optimal point and the optimal value when the objective is to

a maximise $P = 2x + y$,   **b** maximise $P = x + 2y$.

a

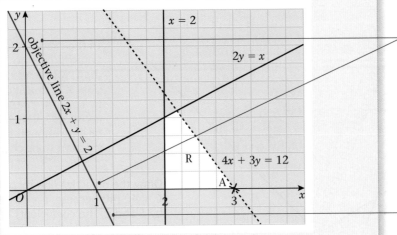

An objective line of the form $P = ax + by$ always passes through $(b, 0)$ and $(0, a)$.

The objective line must be labelled.

Draw an objective line.

The diagram shows the feasible region and the objective line $2x + y = 2$, which passes through $(1, 0)$ and $(0, 2)$.

The final point is point A, where $x = 3$, $y = 0$.

Optimal point is $(3, 0)$.

Optimal value is $P = 2 \times 3 + 0 = 6$.

b

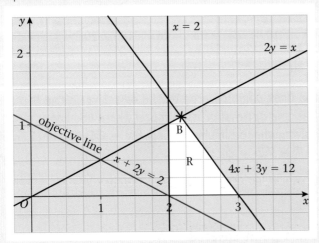

The diagram shows the feasible region and the objective line $x + 2y = 2$, which passes through $(2, 0)$ and $(0, 1)$.

The final point is point B.

B is at the intersection of $2y = x$ and $4x + 3y = 12$.

Solving these simultaneously $11y = 12$, $y = 1\frac{1}{11}$ and $x = 2\frac{2}{11}$.

Optimal point is $(2\frac{2}{11}, 1\frac{1}{11})$.

Optimal value is $P = 2\frac{2}{11} + 2\frac{2}{11} = 4\frac{4}{11}$.

Example 9

In a linear programming problem the constraints are given by

$$3x + y \geqslant 90$$
$$2x + 7y \geqslant 140$$
$$x + y \geqslant 50$$
$$x, y \geqslant 0$$

a minimise $C = 3x + 2y$. **b** minimise $C = 3x + 7y$.

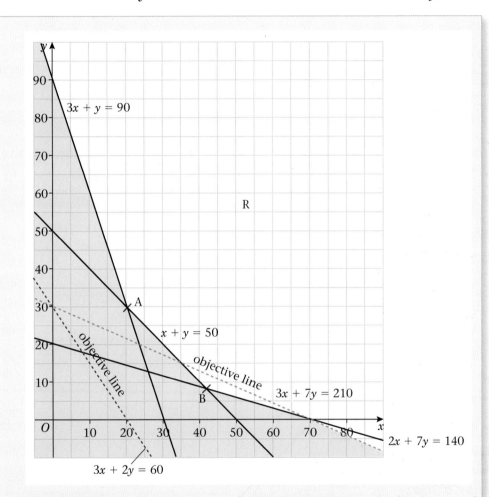

a The objective line has equation $3x + 2y = 60$ and passes through $(20, 0)$ and $(0, 30)$.

The first point in the feasible region as the objective line moves away from the origin is A.

A lies at the intersection of

$$3x + y = 90$$

and $x + y = 50$

Solving simultaneously gives

$$x = 20, y = 30$$

which gives $C = 3 \times 20 + 2 \times 30 = 120$

b The objective line has equation $3x + 7y = 210$ and passes through $(70, 0)$ and $(0, 30)$.

The first point in the feasible region is B.

B lies at the intersection of

$$2x + 7y = 140$$

and $x + y = 50$

Solving simultaneously gives

$$x = 42, y = 8$$

which gives $C = 3 \times 42 + 2 \times 8 = 142$

Example 10

Using the same feasible region as in Example 9, find an optimal solution given that the objective is to minimise $C = x + y$.

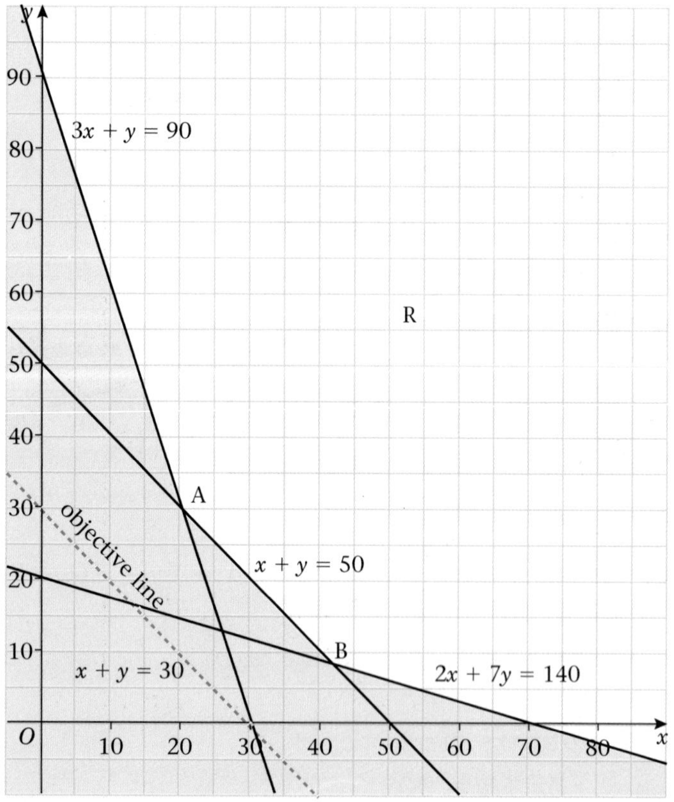

As the objective line slides into the feasible, it lies along the line segment AB. This means that all points along this part of the line are optimal solutions.

For example

A $(20, 30)$ $C = 20 + 30 = 50$

B $(42, 8)$ $C = 42 + 8 = 50$

$(25, 25)$ $C = 25 + 25 = 50$

$(40, 10)$ $C = 40 + 10 = 50$

so any of these points will give an optimal solution. The point $(10, 40)$ also lies on this line, but it is not a solution, since it does not lie in the feasible region.

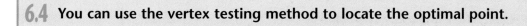

6.4 You can use the vertex testing method to locate the optimal point.

■ You may have noticed that an optimal point will occur at one or more of the vertices of the feasible region. This gives a second method for identifying the optimal point.

1 First find the coordinates of each vertex of the feasible region.

2 Evaluate the objective function at each of these points.

3 Select the vertex that gives the optimal value of the objective function.

Example 11

Use the vertex testing method to find the optimal value of the objective

$$\text{minimise} \quad x + 3y$$
$$\text{subject to} \quad y \leqslant x$$
$$3x + 5y \geqslant 60$$
$$y \geqslant 5$$
$$x \leqslant 13$$
$$x, y \geqslant 0.$$

This is the feasible region determined in Example 7.

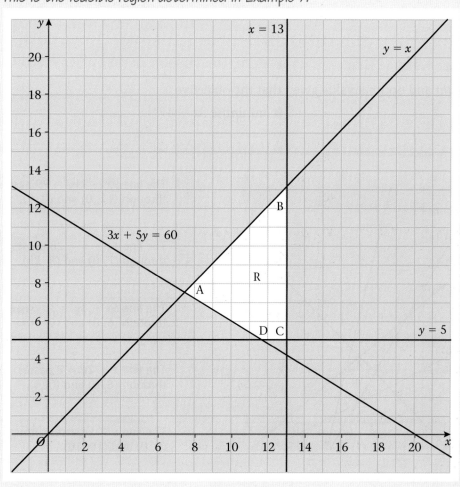

The feasible region has 4 vertices, A, B, C and D.

Vertex	Coordinates	Value of $x + 3y$
A	$x = 7.5 \quad y = 7.5$	$7.5 + 3 \times 7.5 = 30$
B	$x = 13 \quad y = 13$	$13 + 3 \times 13 = 52$
C	$x = 13 \quad y = 5$	$13 + 3 \times 5 = 28$
D	$x = 11\frac{2}{3} \quad y = 5$	$11\frac{2}{3} + 3 \times 5 = 26\frac{2}{3}$

> When using this method all the vertices of the feasible region should be tested.

The minimum value occurs at D $(11\frac{2}{3}, 5)$ and is $26\frac{2}{3}$.

Example 12

A feasible region is defined by the following constraints

$$9x + 11y \leqslant 99$$
$$4x + y \leqslant 28$$
$$5x + 3y \geqslant 30$$
$$x + 2y \geqslant 8$$
$$y \leqslant x$$

Find the optimal point and optimal value given that the objective is

a maximise $x + y$, **b** minimise $3x + 4y$.

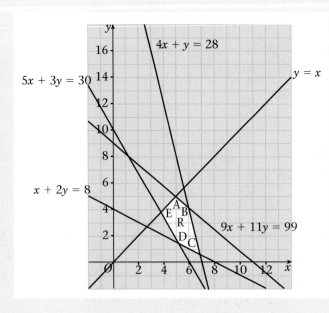

The line $9x + 11y = 99$ passes through $(11, 0)$ and $(0, 9)$.

The line $4x + y = 28$ passes through $(7, 0)$ and $(0, 28)$.

The line $5x + 3y = 30$ passes through $(6, 0)$ and $(0, 10)$.

The line $x + 2y = 8$ passes through $(8, 0)$ and $(0, 4)$.

The line $y = x$ passes through $(0, 0)$ and $(10, 10)$.

The feasible region has 5 vertices, A, B, C, D and E. Use simultaneous equations to find the coordinates of A, B, C, D and E.

Vertex	Coordinates	Value of $x + y$	Value of $3x + 4y$
A	$x = 4\frac{19}{20}$ $y = 4\frac{19}{20}$	$4\frac{19}{20} + 4\frac{19}{20} = 9\frac{9}{10}$	$3 \times 4\frac{19}{20} + 4 \times 4\frac{19}{20} = 34\frac{13}{20}$
B	$x = 5\frac{34}{35}$ $y = 4\frac{4}{35}$	$5\frac{34}{35} + 4\frac{4}{35} = 10\frac{3}{35}$	$3 \times 5\frac{34}{35} + 4 \times 4\frac{4}{35} = 34\frac{13}{35}$
C	$x = 6\frac{6}{7}$ $y = \frac{4}{7}$	$6\frac{6}{7} + \frac{4}{7} = 7\frac{3}{7}$	$3 \times 6\frac{6}{7} + 4 \times \frac{4}{7} = 22\frac{6}{7}$
D	$x = 5\frac{1}{7}$ $y = 1\frac{3}{7}$	$5\frac{1}{7} + 1\frac{3}{7} = 6\frac{4}{7}$	$3 \times 5\frac{1}{7} + 4 \times 1\frac{3}{7} = 21\frac{1}{7}$
E	$x = 3\frac{3}{4}$ $y = 3\frac{3}{4}$	$3\frac{3}{4} + 3\frac{3}{4} = 7\frac{1}{2}$	$3 \times 3\frac{3}{4} + 4 \times 3\frac{3}{4} = 26\frac{1}{4}$

a maximum value of $x + y$ is at B $(5\frac{34}{35}, 4\frac{4}{35})$ and has value of $10\frac{3}{35}$.

b minimum value of $3x + 4y$ is at D $(5\frac{1}{7}, 1\frac{3}{7})$ and has value of $21\frac{1}{7}$.

Exercise 6C

1

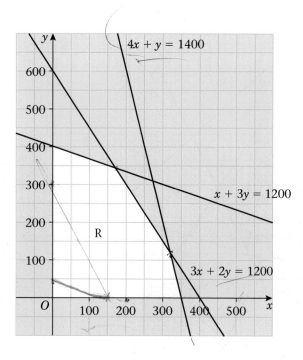

The diagram shows a feasible region, R. Find the optimal point and the optimal value, using
a the objective line method, with the objective 'maximise $M = 2x + y$',
b the objective line method, with the objective 'maximise $N = x + 4y$',
c the vertex testing method, with the objective 'maximise $P = x + y$',
d the vertex testing method, with the objective 'maximise $Q = 6x + y$'.

2

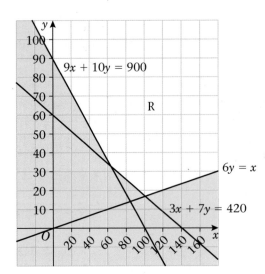

The diagram shows a feasible region, R.

Find the optimal point and the optimal value, using

a the vertex testing method, with the objective 'minimise $E = 2x + y$',

b the vertex testing method, with the objective 'minimise $F = x + 4y$',

c the objective line method, with the objective 'minimise $G = 3x + 4y$',

d the objective line method, with the objective 'minimise $H = x + 6y$'.

3

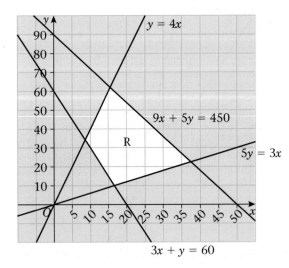

The diagram shows a feasible region, R.

Find the optimal point and the optimal value, using

a the vertex testing method, with the objective 'minimise $J = x + 4y$',

b the vertex testing method, with the objective 'maximise $K = x + y$',

c the objective line method, with the objective 'minimise $L = 6x + y$',

d the objective line method, with the objective 'maximise $M = 2x + y$'.

4

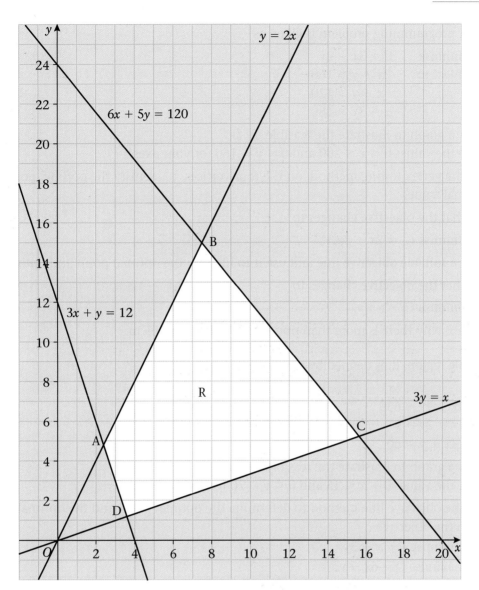

The diagram shows a feasible region, R, which is defined by

$$3x + y \geqslant 12$$
$$y \leqslant 2x$$
$$3y \geqslant x$$
$$6x + 5y \leqslant 120.$$

Determine which vertex, A, B, C or D, is the optimal point for each of the following objectives.

a maximise x

b minimise x

c maximise y

d minimise y

e maximise $6x + y$

f minimise $6x + y$

g maximise $2x + 5y$

h minimise $2x + 5y$

i maximise $3x + 2y$

j minimise $3x + 2y$.

5 A linear programming problem is given as

minimise $C = 3x + 2y$

subject to $2x + y \geqslant 160$

$x + y \geqslant 120$

$x + 3y \geqslant 180$

a Draw a graph to illustrate the feasible, R.

(Take the values $0 \leqslant x \leqslant 200$ and $0 \leqslant y \leqslant 160$ for your axes.)

b Use the vertex testing method, on the four vertices, to identify the optimal point and optimal value.

Given that the objective changes to

minimise $C_1 = 2x + 3y$,

c draw a suitable objective line and use it to identify the optimal point and optimal value.

Given that the objective changes to

minimise $C_2 = x + y$,

d explain why there is more than one solution to the problem.

6 A feasible region, R, is defined by

$y \leqslant 5x$

$14x + 9y \leqslant 630$

$2y \geqslant x$

$4x + y \geqslant 60$

a Draw a graph to illustrate the feasible region, R.

(Take the values $0 \leqslant x \leqslant 45$ and $0 \leqslant y \leqslant 70$ for your axes.)

b Use the objective line method to determine the optimal point and the optimal value given the objective

i minimise $P = x + 3y$,

ii maximise $Q = 6x + y$.

In each case you must draw, and label, an objective line, and find *exact* values.

7 A feasible region, R, is defined by

$y \leqslant 10x$

$x \leqslant 120$

$2y - x \geqslant 100$

$2x + y \leqslant 400$

a Draw a graph to illustrate R.

Given that the objective function is $z = 5x + y$,

b determine the exact value of z, if z is to be

i maximised,

ii minimised.

c Determine the optimal value of $x + 2y$. Give your answer as an exact value.

8 Solve the linear programming problem posed in Exercise 6A, Question 5 (page 120).

9 Solve the linear programming problem posed in Exercise 6A, Question 7 (page 120).

6.5 You can determine solutions that need integer values.

■ There is an additional constraint in some problems – **that the solution has integer values**. For example, questions 1, 2, 3 and 4 in Exercise 6A all require integer solutions, since it is not possible to sell a fraction of a box of chocolates, etc.

■ In such cases the optimal point of the objective may not be an acceptable solution, so you have to find the optimal *integer* solution.

Example 13

Given that x and y must be integers, solve the following linear programming problem.

maximise $\qquad P = x + 2y$

subject to $\qquad 3x + 4y \leqslant 36$

$\qquad\qquad 13x + 9y \leqslant 117$

$\qquad\qquad 5y - 4x \leqslant 10$

$\qquad\qquad 6x + 5y \geqslant 30$

$\qquad\qquad\qquad x \geqslant 2$

Draw a graph of the feasible region.

Show the integer solutions using dots.

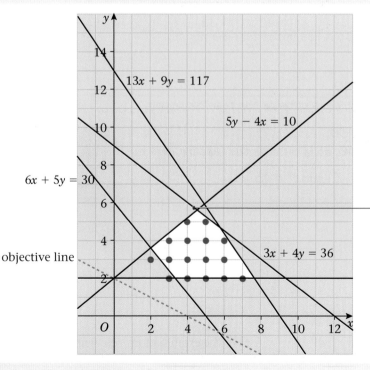

Using the objective line method, the optimal point is at the intersection of $5y - 4x = 10$ and $3x + 4y = 36$, but this does not have integer solutions.

There are two methods that can be used to locate the optimal integer value.

Method 1

If it is possible to plot the integer value solutions accurately, simply select the last integer solution covered by the objective line as it leaves the feasible region, moving away from the origin.

In this case, it is (5, 5) giving $P = 15$.

This method is only possible if the feasible region is sufficiently clear to identify the integer solutions accurately. This may depend on the scales used on the axes.

Method 2

Locate the optimal (non-integer) solution, then test the integer solutions that are close to it. Evaluate the objective function and, most importantly, check that the integer solutions lie in the feasible region.

The optimal solution is $(4\frac{16}{31}, 5\frac{19}{31})$ so test $(4, 5)$, $(4, 6)$, $(5, 5)$ and $(5, 6)$.

It could be that the integer solution lies some distance from the optimal solution. However, in the examination the questions will be such that the optimal integer solution lies close to the optimal solution.

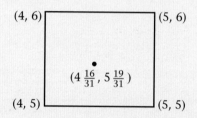

$(4, 5)$ obeys all the constraints, so it lies in the feasible region.

$$P = 4 + 2 \times 5 = 14$$

$(5, 5)$ obeys all the constraints, so it lies in the feasible region.

$$P = 5 + 2 \times 5 = 15$$

$(4, 6)$ does not obey $5y - 4x \leqslant 10$, since $5 \times 6 - 4 \times 4 = 14 \nleqslant 10$, so is outside the feasible region.

$(5, 6)$ does not obey $3x + 4y \leqslant 36$, since $3 \times 5 + 4 \times 6 = 39 \nleqslant 36$, so is outside the feasible region.

The optimal integer solution is $(5, 5)$ and $P = 15$.

Looking at the graph may suggest which constraints may be broken for each point, so these can be checked first.

Example 14

Minimise $x + y$

subject to $3x + 5y \geqslant 1500$

$5x + 2y \geqslant 1000$

$x, y \geqslant 0$

given that x and y must be integers.

In the examination, you would need to decide whether integer solutions were required from the content of the question.

For example, you cannot have fractions of flower displays! (See Exercise 6A Question 2.)

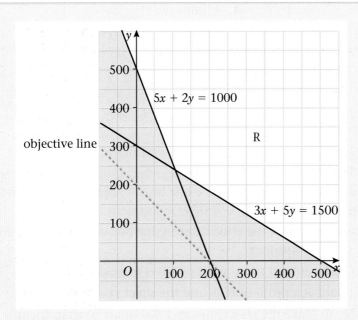

Using the objective line method, the optimal point is at the intersection of $5x + 2y = 1000$

and $3x + 5y = 1500$,

which is $(105\frac{5}{19}, 236\frac{16}{19})$.

You cannot use the objective line to help any more.

Investigate integer solutions close to this point.

> Using a table may help to organise your working.

Point	$5x + 2y \geqslant 1000$	$3x + 5y \geqslant 1500$	In R?	$x + y$
$(105, 236)$	$997 \not\geqslant 1000$	—	No	
$(105, 237)$	$999 \not\geqslant 1000$	—	No	
$(106, 236)$	$1002 \geqslant 1000$	$1498 \not\geqslant 1500$	No	
$(106, 237)$	$1004 \geqslant 1000$	$1503 \geqslant 1500$	Yes	343

> As one constraint has failed, you do not need to check others.

> You only need to evaluate the objective for points in the feasible region.

The optimal integer solution lies at $(106, 237)$ and has a value of 343.

Exercise 6D

Solve the following linear programming problems, given that integer values are required for the decision variables.

1 Maximise $\quad 3x + 2y$

subject to $\quad x + 5y \geqslant 10$

$\qquad\qquad 3x + 4y \leqslant 24$

$\qquad\qquad 4x + 3y \leqslant 24$

$\qquad\qquad x, y \geqslant 0$

2 Minimise $\quad 2x + y$

subject to $\quad 5x + 6y \geqslant 60$

$\qquad\qquad 4x + y \geqslant 28$

$\qquad\qquad x, y \geqslant 0$

3 Maximise $\qquad 5x + 2y$

subject to $\qquad 2y \geqslant x$

$\qquad\qquad 5x + 4y \leqslant 800$

$\qquad\qquad\qquad y \leqslant 4x$

$\qquad\qquad\qquad x, y \geqslant 0$

4 Minimise $\qquad 2x + y$

subject to $\quad 3x + 5y \leqslant 1500$

$\qquad\qquad 3x + 16y \geqslant 2400$

$\qquad\qquad\qquad y \leqslant x$

$\qquad\qquad\qquad x, y \geqslant 0$

Solve these problems from Exercise 6A.

5 A chocolate manufacturer is producing two hand-made assortments, gold and silver, to commemorate 50 years in business.

It will take 30 minutes to make all the chocolates for one box of gold assortment and 20 minutes to make the chocolates for one box of silver assortment.

It will take 12 minutes to wrap and pack the chocolates in one box of gold assortment and 15 minutes for one box of silver assortment.

The manufacturer needs to make at least twice as many silver as gold assortments.

The gold assortment will be sold at a profit of 80p, and the silver at a profit of 60p.

There are 300 hours available to make the chocolates and 200 hours to wrap them.

Maximise the profit, P.

6 A floral display is required for the opening of a new building. The display must be at least 30 m long and is to be made up of two types of planted displays, type A and type B.

Type A is 1 m in length and costs £6

Type B is 1.5 m in length and costs £10

The client wants at least twice as many type A as type B, and at least 6 of type B.

The cost is to be minimised.

7 A toy company makes two types of board game, Cludopoly and Trivscrab. As well as the board each game requires playing pieces and cards.

The company uses two machines, one to produce the pieces and one to produce the cards. Both machines can only be operated for up to ten hours per day.

The first machine takes 5 minutes to produce a set of pieces for Cludopoly and 8 minutes to produce a set of pieces for Trivscrab.

The second machine takes 8 minutes to produce a set of cards for Cludopoly and 4 minutes to produce a set of cards for Trivscrab.

The company knows it will sell at most three times as many games of Cludopoly as Trivscrab.

The profit made on each game of Cludopoly is £1.50 and £2.50 on each game of Trivscrab.

The company wishes to maximise its daily profit.

8 A librarian needs to purchase bookcases for a new library. She has a budget of £3000 and 240 m² of available floor space. There are two types of bookcase, type 1 and type 2, that she is permitted to buy.

Type 1 costs £150, needs 15 m² of floor space and has 40 m of shelving.

Type 2 costs £250, needs 12 m² of floor space and has 60 m of shelving.

She must buy at least 8 type 1 bookcases and wants at most $\frac{1}{3}$ of all the bookcases to be type 2.

She wishes to maximise the total amount of shelving.

Mixed exercise 6E

1 Mr Baker is making cakes and fruit loaves for sale at a charity cake stall. Each cake requires 200 g of flour and 125 g of fruit. Each fruit loaf requires 200 g of flour and 50 g of fruit. He has 2800 g of flour and 1000 g of fruit available.

Let the number of cakes that he makes be x and the number of fruit loaves he makes be y.

a Show that these constraints can be modelled by the inequalities

$x + y \leq 14$ and $5x + 2y \leq 40$.

Each cake takes 50 minutes to cook and each fruit loaf takes 30 minutes to cook. There are 8 hours of cooking time available.

b Obtain a further inequality, other than $x \geq 0$ and $y \geq 0$, which models this time constraint.

c On graph paper illustrate these three inequalities, indicating clearly the feasible region.

d It is decided to sell the cakes for £3.50 each and the fruit loaves for £1.50 each. Assuming that Mr Baker sells all that he makes, write down an expression for the amount of money P, in pounds, raised by the sale of Mr Baker's products.

e Explaining your method clearly, determine how many cakes and how many fruit loaves Mr Baker should make in order to maximise P.

f Write down the greatest value of P.　　　　　　　　　　　　　　　　　　**E**

2 A junior librarian is setting up a music recording lending section to loan CDs and cassette tapes. He has a budget of £420 to spend on storage units to display these items.

Let x be the number of CD storage units and y the number of cassette storage units he plans to buy.

Each type of storage unit occupies $0.08 \, \text{m}^3$, and there is a total area of $6.4 \, \text{m}^3$ available for the display.

a Show that this information can be modelled by the inequality

$x + y \leq 80$

The CD storage units cost £6 each and the cassette storage units cost £4.80 each.

b Write down a second inequality, other than $x \geq 0$ and $y \geq 0$, to model this constraint.

The CD storage unit displays 30 CDs and the cassette storage unit displays 20 cassettes. The chief librarian advises the junior librarian that he should plan to display at least half as many cassettes as CDs.

c Show that this implies that $3x \leq 4y$.

d On graph paper, display your three inequalities, indicating clearly the feasible region.

The librarian wishes to maximise the total number of items, T, on display. Given that

$T = 30x + 20y$

e determine how many CD storage units and how many cassette storage units he should buy, briefly explaining your method.　　　　　　　　　　　　　　　**E**

3 The headteacher of a school needs to hire coaches to transport all the year 7, 8 and 9 pupils to take part in the recording of a children's television programme. There are 408 pupils to be taken and 24 adults will accompany them on the coaches. The headteacher can hire either 54 seater (large) or 24 seater (small) coaches. She needs at least two adults per coach. The bus company has only seven large coaches but an ample supply of small coaches.

Let x and y be the number of large and small coaches hired respectively.

a Show that the situation can be modelled by the three inequalities:
 i $9x + 4y \geqslant 72$
 ii $x + y \leqslant 12$
 iii $x \leqslant 7$.

b On graph paper display the three inequalities, indicating clearly the feasible region.

A large coach costs £336 and a small coach costs £252 to hire.

c Write down an expression, in terms of x and y, for the total cost of hiring the coaches.

d Explain how you would locate the best option for the headteacher, given that she wishes to minimise the total cost.

e Find the number of large and small coaches that the headteacher should hire in order to minimise the total cost and calculate this minimum total cost.

4 The graph below was drawn to solve a linear programming problem. The feasible region, R, includes the points on its boundary.

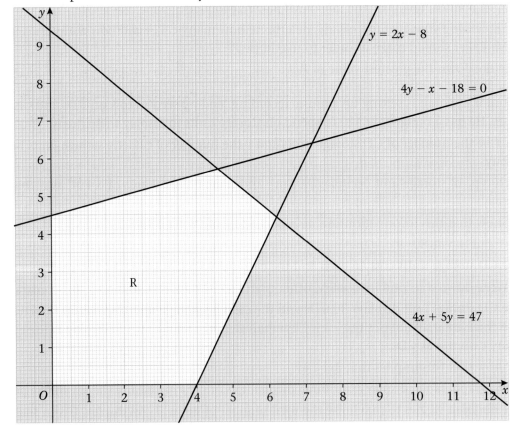

a Write down the inequalities that define the region R.

The objective function, P, is given by $P = 3x + 2y$.

b Find the value of x and the value of y that lead to the maximum value of P. Make your method clear.

c i Give an example of a practical linear programming problem in which it would be necessary for the variables to have integer values.
 ii Given that the solution must have integer values of x and y, find the values that lead to the maximum value of P.

5 A company produces plates and mugs for local souvenir shops. The plates and mugs are manufactured in a two-stage process. Each day there are 300 minutes available for the completion of the first stage and 400 minutes available for the completion of the second stage. In addition the mugs require some hand painting. There are 150 minutes available each day for hand painting.

Product	Stage 1	Stage 2	Hand painting
Plate	$2\frac{1}{2}$	5	—
Mug	3	2	2

The above table shows the production time, in minutes, required for the plates and the mugs.

All plates and mugs made are sold. The profit on each plate sold is £2 and the profit on each mug sold is £4. The company wishes to determine how many plates and mugs to make so as to maximise its profits each day.

Let x be the number of plates made and y the number of mugs made each day.

a Write down the three constraints, other than $x \geqslant 0, y \geqslant 0$, satisfied by x and y.

b Write down the objective function to be maximised.

c Using the graphical method, solve the resulting linear programming problem. Determine the optimal number of plates and mugs to be made each day and the resulting profit.

d When the optimal solution is adopted determine which, if any, of the stages has available time which is unused. State the amount of unused time.

E

Summary of key points

1 To formulate a problem as a linear programming problem:
   - define the decision variables (x, y, z, etc.)
   - state the objective (maximise or minimise, together with an algebraic expression)
   - write the constraints as inequalities.

2 A two-variable linear programming problem can be illustrated graphically.

3 The optimal point in a feasible region can be found using the objective line method or the vertex testing method.

4 Some solutions need integer values.

After studying this chapter you should be able to:

1 model matching problems using a bipartite graph

2 use the maximum matching algorithm.

Matchings

This chapter will show you how to model, and solve, problems involving matching the elements of one set with the elements of another.

This chapter will show you how the staff working in a supermarket could be matched to various tasks, so that each person is given a task that they are qualified to do.

7.1 You need to be able to model matching problems using a bipartite graph.

■ A **bipartite graph** has two sets of nodes.

■ Arcs may connect nodes from different sets but never connect nodes in the same set.

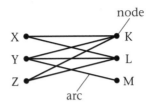

Example 1

The manager of a large department store has five trainees: Robert (R), Stephen (S), Trudy (T), Ursula (U) and Vincent (V), to assign to five departments; Electrical (E), Furnishings (F), Glassware (G), Kitchenware (K), and Lighting (L).

The trainees need to get experience in the departments shown in the table opposite.

Each department must be assigned just one trainee.

Draw a bipartite graph to model this situation.

Trainee	Department
Robert	Electrical, Glassware, Lighting
Stephen	Furnishings, Glassware, Kitchenware
Trudy	Electrical, Furnishings
Ursula	Lighting
Vincent	Furnishings, Kitchenware

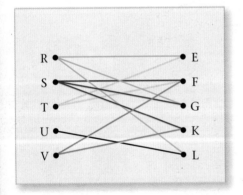

Draw two sets of vertices, one set for the people and the other for the departments. Then use arcs to indicate the department to which each trainee could be assigned.

■ A bipartite graph can be drawn even if the number of nodes in each set is not the same.

Example 2

A gardener has four plots of land, 1, 2, 3 and 4 and needs to decide what to plant in each plot. He can choose from: brassicas (B), flowers (F), peas and beans (P), root crops (R) and shrubs (S). He wishes to grow a different crop on each plot.

Plot 1 is suitable for flowers, peas and beans, and shrubs.
Plot 2 is suitable for brassicas, root crops and shrubs.
Plot 3 is suitable for root crops and shrubs.
Plot 4 is suitable for flowers and brassicas.

Draw a bipartite graph to model this situation.

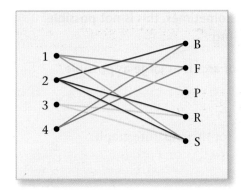

■ Now that you have seen how to use a bipartite graph to model a situation, you now need to look at pairing off some, or all, vertices to form a **matching**.

■ A *matching* is the 1 to 1 pairing of *some*, or *all*, of the elements of one set, X, with elements of a second set, Y.

■ 1 to 1 means that each paired element of set X is paired with just one element of set Y, and each paired element of set Y is paired with just one element of set X.

■ A matching 'pairs off' nodes.

Example 3

Starting from the bipartite graph in Example 1, identify four possible matchings.

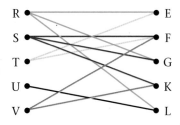

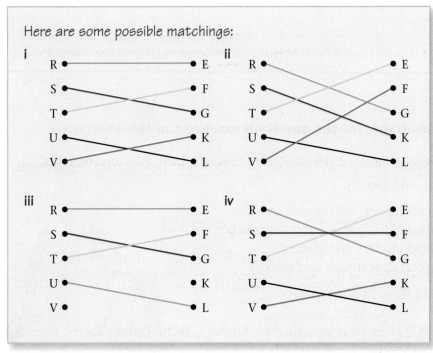

Here are some possible matchings:

There is a *trivial* matching, with no arcs

R •   • E
S •   • F
T •   • G
U •   • K
V •   • L

■ If both sets have *n* nodes, a complete matching is a matching with *n* arcs.

In the solution to Example 3, **i**, **ii** and **iv** are examples of **complete matching**.

■ You will often be asked to find a complete matching but, sometimes, this is not possible. In such cases you will be asked to find a **maximal matching**.

■ A maximal matching is a matching in which the number of arcs is as large as possible.

Example 4

a Explain why it is not possible to find a complete matching for this bipartite graph.

b Find a maximal matching.

a 1 and 3 can only be matched with Y. In a complete matching, two elements in one set cannot be paired with the same element in the other set.

b Since a complete matching, of 4 arcs, is not possible, the best we can find is a matching with 3 arcs. Some possible solutions are:

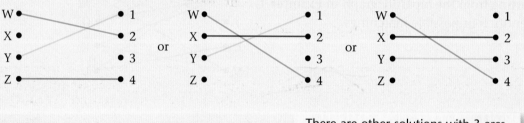

There are other solutions with 3 arcs.

Exercise 7A

Photocopy masters are available for the questions marked * in this exercise.

1 *Four inspectors, Alan, Nicola, Philip and Trudy are to inspect four supermarket departments: bakery, delicatessen, fish and grocery.

Alan is qualified to inspect bakery, delicatessen and fish.
Nicola is qualified to inspect delicatessen, fish and grocery.
Philip is qualified to inspect bakery and grocery.
Trudy is qualified to inspect delicatessen and grocery.

Draw a bipartite graph to show this information.

2 *Tours of the local caves are given daily by six guides: Graham, Keith, Lethna, Preety, Rob, and Vicky. The caves are closed on Wednesdays but guides are needed for Monday, Tuesday, Thursday, Friday, Saturday and Sunday.

Next week the guides are available on the days shown in the following table.

Guide	Days available
Graham	Monday, Friday, Sunday
Keith	Monday, Saturday
Lethna	Tuesday, Thursday
Preety	Friday, Sunday
Rob	Tuesday, Thursday, Sunday
Vicky	Tuesday, Sunday

Represent this information on a bipartite graph.

3 *An athletics team, Bill, Charley, Dara, Eun Jung, Fred and Gopan, will compete in six events: 100 m, 200 m, 1500 m, high jump, long jump and javelin. Each person may only compete in one event.

Bill prefers the high jump or 200 m.
Charley does not like the running events, but is happy to compete in either of the jumping events or the javelin.
Dara will do the 100 m, 200 m or the high jump.
Eun Jung likes competing in 200 m, 1500 m or javelin.
Fred prefers long jump or 100 m.
Gopan will do any of the running events 100 m, 200 m or 1500 m.

Draw a bipartite graph to represent this information.

7.2 You can use the maximum matching algorithm, starting from an initial matching.

■ When you have drawn a bipartite graph, you find an initial matching, then apply the **maximum matching algorithm** to find an improved matching.

> In the examination you will be given an initial matching from which you must start.

■ An **alternating path** starts at an unmatched node on one side of the bipartite graph and finishes at an unmatched node on the other side. It uses arcs that are alternately 'not in' or 'in' the initial matching.

■ The alternating path must take into account the information given. For example, when matching people to tasks, an alternating path can only match people to the tasks they are prepared to do.

■ Here is the maximum matching algorithm.

> The maximum matching may not be *unique*. If there is more than one alternating path possible, you may be able to find more than one maximum matching.

1 Start with any (non-trivial) initial matching.

2 Search for an alternating path.

3 If an alternating path can be found, use it to create an improved matching by changing the status of the arcs. If an alternating path cannot be found, stop.

4 List the new matching, which consists of the result of applying the alternating path together with any unchanged elements of the initial matching.

5 If the matching is now complete, stop. Otherwise return to step 2.

Example 5

At the County Fair, five judges, Pat, Ramin, Sze Ting, Tom and Will, are to judge five events: cakes, flowers, fruit, preserves and vegetables.

Pat judged flowers last year but could judge fruit or cakes.
Ramin judged cakes last year but could judge preserves.
Sze Ting judged fruit last year but could judge cakes.
Tom is a new judge and can only judge fruit.
Will judged preserves last year but could judge flowers or vegetables.

a Draw a bipartite graph to illustrate this information.

Initially Pat, Ramin, Sze Ting and Will are matched to the event they judged last year.

> This is the initial matching you are to use.

b Show this initial matching on a bipartite graph.

c Starting from the initial matching, use the maximum matching algorithm to find a complete matching. List any alternating paths you use.

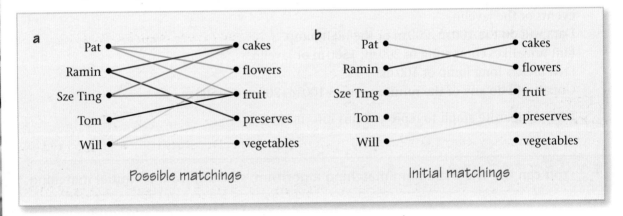

Now apply the maximum matching algorithm. You need to find an alternating path.

Looking at the initial matching diagram, there is only one unmatched vertex on each side. You need an alternating path from 'Tom' to 'vegetables'.

The arcs must alternate, starting with an arc that is not in the matching at the moment.

Alternating path

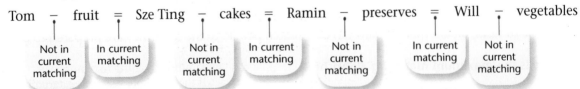

This is an alternating path.
A breakthrough has occurred.

> You can 'read' this path as − meaning 'could do' and = meaning 'currently being done by'

Now **change status**.

> 'Change status', simply means switch the symbols over. Switching the 'in' and 'not in' edges.

Tom = fruit − Sze Ting = cakes − Ramin = preserves − Will = vegetables

Now list the new matching.

We write down Tom's, Sze Ting's, Ramin's and Will's new matchings first.

It is simplest to list the judges first, then look at the changed status alternating path and implement these changes next. Other matchings remain unchanged.

Final matching

Pat = flowers

Pat's matching was unaffected by the alternating path and so remains unchanged.

Ramin = preserves

Sze Ting = cakes

Tom = fruit

Will = vegetables

This is a complete matching, so stop.

Each time you apply the maximum matching algorithm and find an alternating path, you increase the number of matched pairs by one.

■ **If there is more than one pair of unmatched nodes you will need to find more than one alternating path.**

■ **Each subsequent application of the algorithm requires you to start from the current matching.**

Example 6

A college dramatic society has six helpers: Andrew (A), Deepa (D), Henry (H), Karl (K), Nicola (N) and Yi-Ying (Y),

They are to be matched to six tasks: props (P), lighting (L), make-up (M), sound (S), tickets (T) and wardrobe (W).

This table indicates which tasks each person is able to do.

Name	Tasks
Andrew	wardrobe, props, tickets
Deepa	tickets, make-up
Henry	lighting, make-up
Karl	sound, wardrobe, lighting
Nicola	sound
Yi-Ying	lighting, tickets

a Draw a bipartite graph to model this situation.
   Initially Andrew, Deepa, Henry and Karl are matched to the first task on their individual lists.

b Indicate this initial matching on a new bipartite graph.

c Starting from this matching, use the maximum matching algorithm to find a complete matching. Indicate clearly how the algorithm has been applied, listing any alternating paths.

a
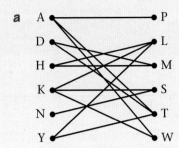

b
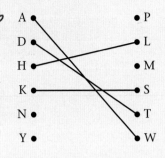

c
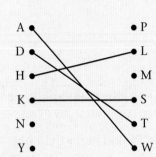

In order to make the initial matching distinctive, a wavy line has been used. You must *not* use colour in the examination.

There are *two* unmatched nodes on the left, N and Y, and *two* unmatched nodes on the right, P and M.

A complete matching will need *two* alternating paths.

The alternating paths must take into account the tasks people are prepared to do. (See table on page 155.)

There is a choice for a starting node. Choose N. Look for an alternating path from N to P or M.

The alternating path must alternate between arcs that are 'not in' and 'in' the current matching.

The alternating path must begin.

$$N - S = K ...$$

not in current

in current

Nicola is only prepared to do sound.

Then there is a choice between K − L or K − W.

A decision tree diagram can be used to explore the choices.

Karl does not do make-up or props!

$$N - S = K \begin{cases} L = H - M \\ W = A - P \end{cases}$$

A breakthrough has occurred.

A breakthrough has occurred.

We choose the top of these two possible alternating paths.

Change the status N = S − K = L − H = M.

The matching is now
A = W
D = T
H = M
K = L
N = S
Y

Here is the new bipartite graph.

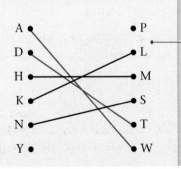

The blue lines show those matchings that are still valid from the initial matching.

You do not need to redraw the graph each time, as long as you use the updated matching.

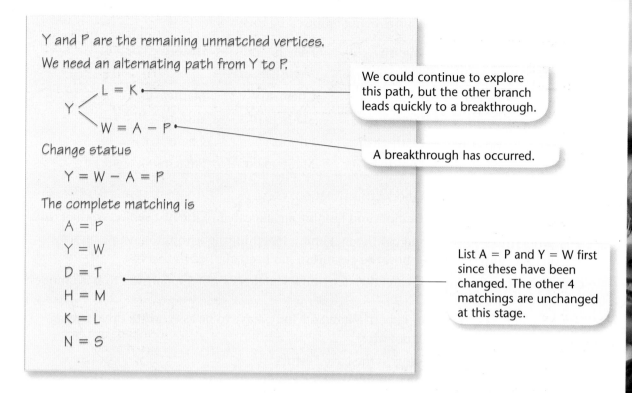

Y and P are the remaining unmatched vertices.
We need an alternating path from Y to P.

L = K •————————————
Y <
W = A – P •————

We could continue to explore this path, but the other branch leads quickly to a breakthrough.

Change status

Y = W – A = P

A breakthrough has occurred.

The complete matching is

A = P
Y = W
D = T •————————————
H = M
K = L
N = S

List A = P and Y = W first since these have been changed. The other 4 matchings are unchanged at this stage.

Mixed exercise 7B

Photocopy masters are available for the questions marked * in this exercise.

1 *The tour director of a museum needs to allocate five of his guides to parties of tourists from France, Germany, Italy, Japan and Spain. The table shows the languages spoken by the five guides.

Ruth	French	Spanish	
Steve	German	Japanese	
Tony	French	German	
Ursula	Spanish	Italian	
Victoria	Italian	Spanish	Japanese

a Draw a bipartite graph to model this situation.

The director allocates Ruth, Steve, Ursula and Victoria to the parties who speak the first language in their individual lists.

b Starting from this matching use the maximum matching algorithm to find a complete matching. Indicate clearly how the algorithm has been applied in this case. **E**

2 *In order to help new A-level students to select their courses a college organises an open evening. Some students already studying A-level courses have agreed to talk about one of their A-level courses. Six of these students, Ann, Barry, David, Gemma, Jasmine and Nickos, are, between them, following six A-level courses in Chemistry, English, French, History, Mathematics and Physics.

The table below shows the courses being followed by each student:

Ann	English	French	History
Barry	History	English	
David	French	Chemistry	Mathematics
Gemma	Physics	Mathematics	History
Jasmine	Mathematics	Physics	Chemistry
Nickos	English	Mathematics	French

a Draw a bipartite graph to model this situation.

Initially Ann, Barry, David, Gemma and Jasmine are allocated to the first subject in their lists.

b Starting from this matching use the maximum matching algorithm to find a complete matching. Indicate clearly how the algorithm has been applied in this case.

c Explaining your reasoning carefully, determine whether or not your answer to **b** is unique.

3 * Five coach drivers, Mihi, Pat, Robert, Sarah and Tony, have to be assigned to drive five coaches for the following school trips:

Adupgud Senior School is going to the Lake District
Brayknee Junior School is going to the seaside
Korry Stur Junior School is going to a concert
Learnalott Senior School is going to the museum (two coaches needed)

Mihi and Sarah would like to drive senior school children. Robert and Pat would like to go on the seaside trip. Pat and Tony would like to attend the concert. Robert and Pat would like to visit the museum. Pat and Tony would like to visit the Lake District.

The driver manager wishes to assign each driver to a trip they would like to do.

a Draw a bipartite graph to show the trips that the drivers would like to take.

Initially Mihi and Pat are assigned to Learnalott Senior School, Robert is assigned to Brayknee Junior School and Tony is assigned to Adupgud Senior School.

b Starting from this matching use the maximum matching algorithm to find a complete matching. You must indicate clearly how the algorithm has been applied in this case. State your alternating path and the final matching.

E

4

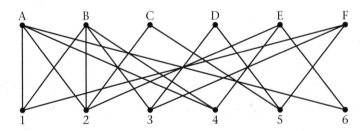

The bipartite graph above shows a mapping between six volunteers, A, B, C, D, E and F, and six tasks, 1, 2, 3, 4, 5 and 6. The lines indicate which tasks each volunteer is qualified to do.

The initial matching is A − 2, B − 1, C − 5, D − 3 and E − 4.

a Starting from this matching, use the maximum matching algorithm to find a complete matching. You must indicate clearly how the algorithm has been applied in this case. State your alternating path and your final matching.

Volunteer E now insists on doing task 2.

b State the changes that need to be made to the initial model to accommodate this.

5 * At a school fete six teachers, A, B, C, D, E and F, are to run six stalls, R, S, T, U, V and W.

A would prefer to run T but is willing to run R
B would prefer to run S but is willing to run R or W
C would prefer to run U but is willing to run S
D would prefer to run V but is willing to run R
E is willing to run T or V
F is willing to run V

a Draw a bipartite graph to model this situation.

Initially, A, B, C and D are matched to their preferred choices.

b Indicate this initial matching in a distinctive way on the bipartite graph drawn in **a**.

c Use the maximum matching algorithm to find a maximum matching, listing clearly your alternating path.

d Explain why it is not possible to find a complete matching. You should make specific reference to individual stalls and teachers.

Teacher A now offers to run stall S.

e Draw a new bipartite graph. Hence, using the previous initial matching and the maximum matching algorithm, determine if it is now possible to obtain a complete matching. If it is possible, give the matching, stating clearly your alternating path; if it is still not possible explain why.

6

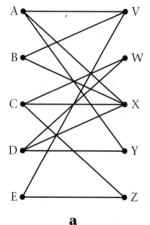

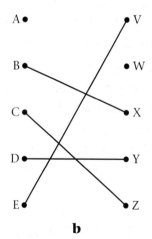

a   **b**

Five people A, B, C, D and E are to be matched to five tasks V, W, X, Y and Z. A bipartite graph showing the possible matchings is shown in **a**, and an initial matching is shown in **b**. There are a number of distinct alternating paths that can be generated from this initial matching. Two such paths are

A − Y − D − W and A − X − B − V − E − Z − C − W.

a Use each of these two alternating paths, in turn, to write down the complete matchings they generate.

b Using the maximum matching algorithm and the given initial matching,

  i   find two further distinct alternating paths, making your reasoning clear,

  ii write down the complete matchings they generate.

7 * A college has five vacant jobs, A, B, C, D and E. There are five applicants who are labelled 1, 2, 3, 4 and 5. The applicants are only qualified for certain jobs and the following table summarises this information.

The diagram shows a bipartite graph modelling this information.

Applicant	Jobs qualified for
1	C, D
2	B, D, E
3	A, C, E
4	A, C
5	A, D

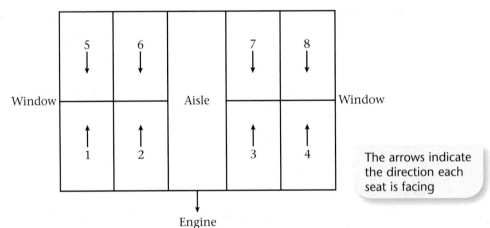

Initially Applicant 2 is allocated to job B, 3 is attached to C and 5 is allocated to A.

a i Show this matching clearly on a diagram.

ii Starting from this matching, use the maximum matching algorithm to obtain an improved matching. State clearly your alternating path and show this improved matching on a diagram.

iii Hence obtain a complete matching. State clearly your alternating path and this complete matching.

The interviewing committee decides that applicant 3 is to be appointed to job C.

b If this appointment is made, explain why it is not possible to fill all the other jobs with the remaining applicants.

E (adapted)

8 *

The diagram above represents eight seats in a railway carriage, numbered 1, 2, 3, 4, 5, 6, 7 and 8. These are the last eight seats available on a special sightseeing trip. The booking clerk has to arrange the seating for the final customers. Six customers make the following requests:

Ms A wants an aisle seat facing the engine (6 or 7)

Mr B wants a window seat (1, 4, 5 or 8)

Rev C wants a seat with his back to the engine (1, 2, 3 or 4)

Mrs D wants an aisle seat (2, 3, 6 or 7)

Miss E wants a seat facing the engine (5, 6, 7 or 8)

Dr F wants a window seat with her back to the engine (1 or 4)

Initially the clerk assigns the seats as follows:

A to 6, B to 5, C to 4, D to 2, E to 7 and F to 1

The day before departure Mr and Mrs G join the trip. They ask to sit next to each other (1 and 2, 3 and 4, 5 and 6 or 7 and 8). The clerk reassigns the seats, using as far as possible the original seat assignments as the initial matching.

a Choose two seats for Mr and Mrs G and, using a bipartite graph, model the possible seat allocations of the other customers.

b Indicate, in a distinctive way, those elements of the clerk's original matching that are still possible.

c Using your answer to part **b** as the initial matching, apply the maximum matching algorithm. You must state your alternating path and your final maximum matching.

Summary of key points

1 A **bipartite graph** has two sets of nodes.

2 Arcs may connect nodes from different sets but never connect nodes in the same set.

3 An **alternating path** starts at an unmatched node on one side of a bipartite graph and finishes at an unmatched node on the other side. It uses arcs that are alternately 'not in' and 'in' the initial matching.

4 When you have drawn a bipartite graph, you can find an improved matching using the **maximum matching algorithm**.

5 The maximum matching algorithm is as follows.
- Start with any initial matching.
- Search for an alternating path.
- If an alternating path can be found, use it to create an improved matching by changing the status of the arcs (in → out, out → in). If an alternating path cannot be found, stop.
- List the new matching, which consists of the result of applying the alternating path together with any unchanged elements of the previous matching.
- If the matching is now complete, stop. Otherwise look for another alternating path.

6 If there is more than one pair of unmatched nodes you will need to find more than one alternating path.

7 Each time you apply the maximum matching algorithm you increase the number of matched pairs by one.

Review Exercise

Photocopy masters are available for the questions marked * in this exercise.

1 Define the terms

  a bipartite graph,

  b alternating path,

  c matching,

  d complete matching. **E**

2* The organiser of a sponsored walk wishes to allocate each of six volunteers, Alan, Geoff, Laura, Nicola, Philip and Sam to one of the checkpoints along the route. Two volunteers are needed at checkpoint 1 (the start) and one volunteer at each of checkpoints 2, 3, 4 and 5 (the finish). Each volunteer will be assigned to just one checkpoint. The table shows the checkpoints each volunteer is prepared to supervise.

Name	Checkpoints
Alan	1 or 3
Geoff	1 or 5
Laura	2, 1 or 4
Nicola	5
Philip	2 or 5
Sam	2

Initially Alan, Geoff, Laura and Nicola are assigned to the first checkpoint in their individual list.

  a Draw a bipartite graph to model this situation and indicate the initial matching in a distinctive way.

  b Starting from this initial matching, use the maximum matching algorithm to find an improved matching. Clearly list any alternating paths you use.

  c Explain why it is not possible to find a complete matching. **E**

3* Ann, Bryn, Daljit, Gareth and Nickos have all joined a new committee. Each of them is to be allocated to one of five jobs 1, 2, 3, 4 or 5. The table shows each member's preferences for the jobs.

Ann	1 or 2
Bryn	3 or 1
Daljit	2 or 4
Gareth	5 or 3
Nickos	1 or 2

Initially Ann, Bryn, Daljit and Gareth are allocated the first job in their lists shown in the table.

a Draw a bipartite graph to model the preferences shown in the table and indicate, in a distinctive way, the initial allocation of jobs.

b Use the matching improvement algorithm to find a complete matching, showing clearly your alternating path.

c Find a second alternating path from the initial allocation. **E**

4 The precedence table for activities involved in a small project is shown below.

Activity	Preceding activities
A	–
B	–
C	–
D	B
E	A
F	A
G	B
H	C, D
I	E
J	E
K	F, G, I
L	H, J, K

Draw an activity network, using activity on edge and without using dummies, to model this project. **E**

5 The precedence table for activities involved in manufacturing a toy is shown below.

Activity	Preceding activities
A	–
B	–
C	–
D	A
E	A
F	B
G	B
H	C, D, E, F
I	E
J	E
K	I
L	I
M	G, H, K

a Draw an activity network, using activity on arc, and exactly one dummy, to model the manufacturing process.

b Explain briefly why it is necessary to use a dummy in this case. **E**

6 *

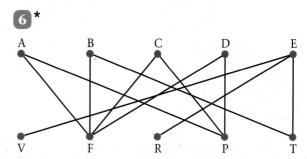

The five winners of a competition are Mr. Adams (A), Mr. Brown (B), Miss Church (C), Mrs. Drain (D) and Ms. Eagle (E). The prizes are five cars; a Vauxhall (V), a Ford (F), a Rover (R), a Peugeot (P) and a Toyota (T). The winners' preferences are summarised in the bipartite graph G shown above.

The organiser of the competition matches C with F, D with P and E with T.

a Indicate clearly this matching M on a matching diagram.

b Find an alternating path for M in G, starting at B. Use this to construct an improved matching M′. Show M′ on another diagram.

c Show that there is no alternating path for M′ in G. **E**

7 * At Tesafe supermarket there are 5 trainee staff, Homan (H), Jenna (J), Mary (M), Tim (T) and Yoshie (Y). They each must spend one week in each of 5 departments, Delicatessen (D), Frozen foods (F), Groceries (G), Pet foods (P), Soft drinks (S). Next week every department requires exactly one trainee. The table below shows the departments in which the trainees have yet to spend time.

Trainee	Departments
H	D, F, P
J	G, D, F
M	S, P, G
T	F, S, G
Y	D

Initially H, J, M and T are allocated to the first department in their list.

a Draw a bipartite graph to model this situation and indicate the initial matching in a distinctive way.

Starting from this matching,

b use the maximum matching algorithm to find a complete matching. You must make clear your alternating path and your complete matching. **(E)**

8 a Draw an activity network for the project described in this precedence table, using as few dummies as possible.

Activity	Must be preceded by:
A	–
B	A
C	A
D	A
E	C
F	C
G	B, D, E, F
H	B, D, E, F
I	F, D
J	G, H, I
K	F, D
L	K

b A different project is represented by the activity network shown below. The duration of each activity is shown in brackets.

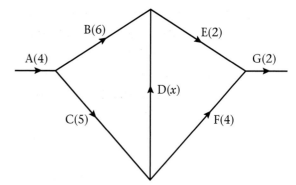

Find the range of values of x that will make D a critical activity. **(E)**

9

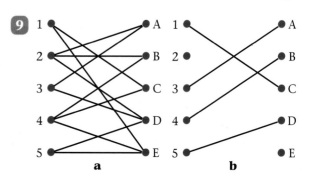

Five members of staff 1, 2, 3, 4 and 5 are to be matched to five jobs A, B, C, D and E. A bipartite graph showing the possible matchings is given in **a** and an initial matching M is given in **b**.

There are several distinct alternating paths that can be generated from M. Two such paths are

$$2 - B = 4 - E$$

and $2 - A = 3 - D = 5 - E.$

a Use each of these two alternating paths, in turn, to write down the complete matchings they generate.

Use the maximum matching algorithm and the initial matching M,

b find two further distinct alternating paths, making your reasoning clear. **(E)**

10 * Two fertilisers are available, a liquid X and a powder Y. A bottle of X contains 5 units of chemical A, 2 units of chemical B and $\frac{1}{2}$ unit of chemical C. A packet of Y contains 1 unit of A, 2 units of B and 2 units of C. A professional gardener makes her own fertiliser. She requires at least 10 units of *A*, at least 12 units of *B* and at least 6 units of *C*.

She buys x bottles of *X* and y packets of *Y*.

a Write down the inequalities which model this situation.

b On the grid provided construct and label the feasible region.

A bottle of *X* costs £2 and a packet of *Y* costs £3.

c Write down an expression, in terms of x and y, for the total cost £*T*.

d Using your graph, obtain the values of x and y that give the minimum value of *T*. Make your method clear and calculate the minimum value of *T*.

e Suggest how the situation might be changed so that it could no longer be represented graphically. **E**

11 * At a water sports centre there are five new instructors, Ali (A), George (G), Jo (J), Lydia (L) and Nadia (N). They are to be matched to five sports, canoeing (C), scuba diving (D), surfing (F), sailing (S) and water skiing (W).

The table indicates the sports each new instructor is qualified to teach.

Instructor	Sport
A	C, F, W
G	F
J	D, C, S
L	S, W
N	D, F

Initially, A, G, J and L are each matched to the first sport in their individual list.

a Draw a bipartite graph to model this situation and indicate the initial matching in a distinctive way.

b Starting from this initial matching, use the maximum matching algorithm to find a complete matching. You must clearly list any alternating paths used.

Given that on a particular day J must be matched to D,

c explain why it is no longer possible to find a complete matching. **E**

12 A company produces two types of self-assembly wooden bedroom suites, the 'Oxford' and the 'York'. After the pieces of wood have been cut and finished, all the materials have to be packaged. The table below shows the time, in hours, needed to complete each stage of the process and the profit made, in pounds, on each type of suite.

	Oxford	York
Cutting	4	6
Finishing	3.5	4
Packaging	2	4
Profit (£)	300	500

The times available each week for cutting, finishing and packaging are 66, 56 and 40 hours respectively.

The company wishes to maximise its profit.

Let x be the number of Oxford, and y be the number of York suites made each week.

a Write down the objective function.

b In addition to
$$2x + 3y \leqslant 33,$$
$$x \geqslant 0,$$
$$y \geqslant 0,$$
find two further inequalities to model the company's situation.

c Illustrate all the inequalities, indicating clearly the feasible region.

d Explain how you would locate the optimal point.

e Determine the number of Oxford and York suites that should be made each week and the maximum profit gained.

It is noticed that when the optimal solution is adopted, the time needed for one of the three stages of the process is less than that available.

f Identify this stage and state by how many hours the time may be reduced. **E**

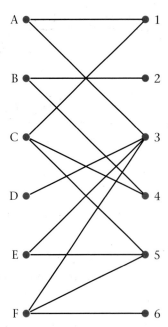

13 *

The bipartite graph above shows the possible allocations of people A, B, C, D, E and F to tasks 1, 2, 3, 4, 5 and 6.

An initial matching is obtained by matching the following pairs

A to 3,   B to 4,   C to 1,   F to 5.

a Show this matching in a distinctive way on a diagram.

b Use an appropriate algorithm to find a maximal matching. You should state any alternating paths you have used. **E**

14 * Six newspaper reporters Asif (A), Becky (B), Chris (C), David (D), Emma (E) and Fred (F), are to be assigned to six news stories Business (1), Crime (2), Financial (3), Foreign (4), Local (5) and Sport (6). The table shows possible allocations of reporters to news stories. For example, Chris can be assigned to any one of stories 1, 2 or 4.

	1	2	3	4	5	6
A					✔	
B	✔			✔		
C	✔	✔		✔		
D					✔	
E			✔		✔	✔
F				✔		

a Show these possible allocations on a bipartite graph.

A possible matching is

A to 5,   C to 1,   E to 6,   F to 4.

b Show this information, in a distinctive way, on a diagram.

c Use an appropriate algorithm to find a maximal matching. You should list any alternating paths you have used.

d Explain why it is not possible to find a complete matching. **E**

15 *

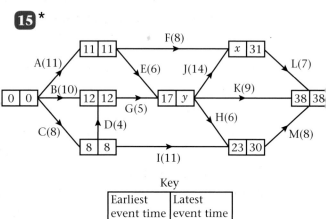

F(8)

Key

Earliest event time	Latest event time

A project is modelled by the activity network in the diagram. The activities are represented by the arcs. One worker is required for each activity. The number in brackets on each arc gives the time, in hours, to complete the activity. The earliest event time and the latest event time are given by the numbers in the left box and right box respectively.

a State the value of x and the value of y.

b List the critical activities.

c Explain why at least 3 workers will be needed to complete this project in 38 hours.

d Schedule the activities so that the project is completed in 38 hours using just 3 workers. You must make clear the start time and finish time of each activity. **E**

16 * The network opposite shows the activities involved in the process of producing a perfume. The activities are represented by the arcs. The number in brackets on each arc gives the time, in hours, taken to complete the activity.

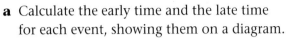

C(23)   F(10)
A(12)   J(6)   K(19)
D(14)   H(18)   L(13)
G(15)
B(17)   E(32)   M(27)
I(20)

a Calculate the early time and the late time for each event, showing them on a diagram.

b Hence determine the critical activities.

c Calculate the total float time for D.

Each activity requires only one person.

d Find a lower bound for the number of workers needed to complete the process in the minimum time.

Given that there are only three workers available, and that workers may **not** share an activity,

e schedule the activities so that the process is completed in the shortest time. Use a time line. State the new shortest time. **E**

17 The Young Enterprise Company 'Decide', is going to produce badges to sell to decision maths students. It will produce two types of badge.

Badge 1 reads 'I made the decision to do maths' and

Badge 2 reads 'Maths is the right decision'

'Decide' must produce at least 200 badges and has enough material for 500 badges.

Market research suggests that the number produced of Badge 1 should be between 20% and 40% of the total number of badges made.

The company makes a profit of 30p on each Badge 1 sold and 40p on each Badge 2. It will sell all that it produces, and wishes to maximise its profit.

Let x be the number produced of Badge 1 and y be the number of Badge 2.

a Formulate this situation as a linear programming problem, simplifying your inequalities so that all the coefficients are integers.

b On suitable axes, construct and clearly label the feasible region.

c Using your graph, advise the company on the numbers of each badge it should produce. State the maximum profit 'Decide' will make. **E**

18 * A building project is modelled by the activity network shown opposite. The activities are represented by the arcs. The number in brackets on each arc gives the time, in hours, taken to complete the activity. The left box entry at each vertex is the earliest event time and the right box entry is the latest event time

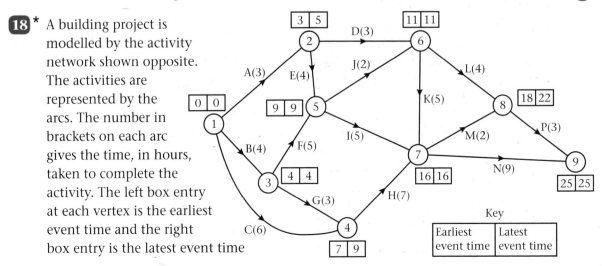

Key

Earliest event time	Latest event time

a Determine the critical activities and state the length of the critical path.

b State the total float for each non-critical activity.

c On a grid, draw a cascade (Gantt) chart for the project.

Given that each activity requires one worker,

d draw up a schedule to determine the minimum number of workers required to complete the project in the critical time. State the minimum number of workers. **E**

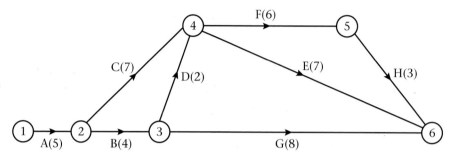

A project is modelled by the activity network shown above. The activities are represented by the edges. The number in brackets on each edge gives the time, in days, taken to complete the activity.

a Calculate the early time and the late time for each event. Write these in the boxes on the worksheet.

b Hence determine the critical activities and the length of the critical path.

c Obtain the total float for each of the non-critical activities.

d Draw a cascade (Gantt) chart showing the information obtained in parts **b** and **c**.

Each activity requires one worker. Only two workers are available.

e Draw up a schedule and find the minimum time in which the 2 workers can complete the project. **E**

20 Becky's bird food company makes two types of bird food. One type is for bird feeders and the other for bird tables. Let x represent the quantity of food made for bird feeders and y represent the quantity of food made for bird tables. Due to restrictions in the production process, and known demand, the following constraints apply.

$$x + y \leqslant 12,$$
$$y < 2x,$$
$$2y \geqslant 7,$$
$$y + 3x \geqslant 15.$$

a Show these constraints on a diagram and label the feasible region R.

The objective is to minimise $C = 2x + 5y$.

b Solve this problem, making your method clear. Give, as fractions, the value of C and the amount of each type of food that should be produced.

Another objective (for the same constraints given above) is to maximise $P = 3x + 2y$, where the variables must take integer values.

c Solve this problem, making your method clear. State the value of P and the amount of each type of food that should be produced. **E**

21 *

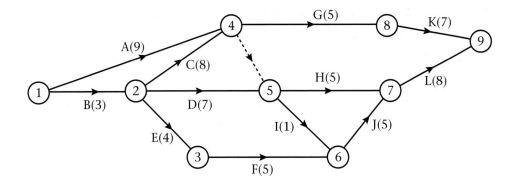

A project is modelled by the activity network shown above. The activities are represented by the arcs. The number in brackets on each arc gives the time, in hours, to complete the activity. The numbers in circles give the event numbers. Each activity requires one worker.

a Explain the purpose of the dotted line from event 4 to event 5.

b Calculate the early time and the late time for each event.

c Determine the critical activities.

d Obtain the total float for each of the non-critical activities.

e On a grid, draw a cascade (Gantt) chart, showing the answers to parts **c** and **d**.

f Determine the minimum number of workers needed to complete the project in the minimum time. Make your reasoning clear. **E**

22 The company EXYCEL makes two types of battery, X and Y. Machinery, workforce and predicted sales determine the number of batteries EXYCEL make. The company decides to use a graphical method to find its optimal daily production of X and Y.

The constraints are modelled in the diagram opposite.

The feasible region, R, is indicated.

x = the number (in thousands) of type X batteries produced each day,

y = the number (in thousands) of type Y batteries produced each day.

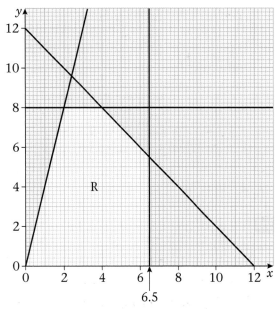

The profit of each type X battery is 40p and on each type Y battery is 20p.

The company wishes to maximise its daily profit.

a Write this as a linear programming problem, in terms of x and y, stating the objective function and all the constraints.

b Find the optimal number of batteries to be made each day. Show your method clearly.

c Find the daily profit, in £, made by EXYCEL. **E**

23 *

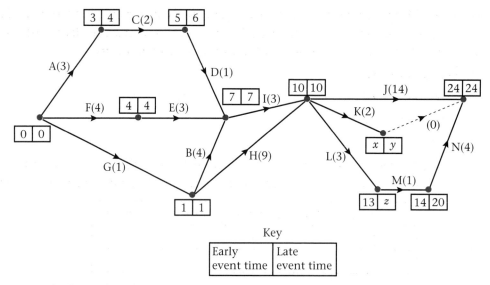

Key

Early event time	Late event time

The network above shows activities that need to be undertaken in order to complete a project. Each activity is represented by an arc. The number in brackets is the duration of the activity in hours. The early and late event times are shown at each node. The project can be completed in 24 hours.

a Find the values of x, y and z.

b Explain the use of the dummy activity in the diagram.

c List the critical activities.

d Explain what effect a delay of one hour to activity B would have on the time taken to complete the whole project.

The company which is to undertake this project has only two full-time workers available. The project must be completed in 24 hours and in order to achieve this, the company is prepared to hire additional workers at a cost of £28 per hour. The company wishes to minimise the money spent on additional workers. Any worker can undertake any task and each task requires only one worker.

e Explain why the company will have to hire additional workers in order to complete the project in 24 hours.

f Schedule the tasks to workers so that the project is completed in 24 hours and at minimum cost to the company.

g State the minimum extra cost to the company.   **E**

24 *

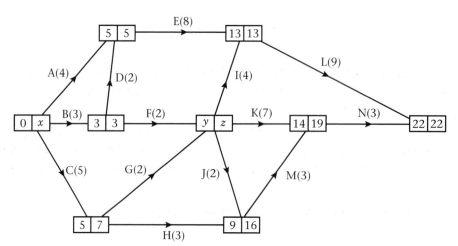

A trainee at a building company is using critical path analysis to help plan a project. The diagram shows the trainee's activity network. Each activity is represented by an arc and the number in brackets on each arc is the duration of the activity, in hours.

a Find the values of x, y and z.

b State the total length of the project and list the critical activities.

c Calculate the total float time on

    i activity N,   **ii** activity H.

The trainee's activity network is checked by the supervisor who finds a number of errors and omissions in the diagram. The project should be represented by the following precedence table.

Activity	Must be preceded by:	Duration
A	–	4
B	–	3
C	–	5
D	B	2
E	A, D	8
F	B	2
G	C	2
H	C	3
I	F, G	4
J	F, G	2
K	F, G	7
L	E, I	9
M	H, J	3
N	E, I, K, M	3
P	E, I	6
Q	H, J	5
R	Q	7

d By adding activities and dummies amend the diagram so that it represents this precedence table.

e Find the total time needed to compete this project.   **E**

25 A chemical company produces two products X and Y. Based on potential demand, the total production each week must be at least 380 gallons. A major customer's weekly order for 125 gallons of Y must be satisfied.

Product X requires 2 hours of processing time for each gallon and product Y requires 4 hours of processing time for each gallon. There are 1200 hours of processing time available each week. Let x be the number of gallons of X produced and y be the number of gallons of Y produced each week.

a Write down the inequalities which x and y must satisfy.

It costs £3 to produce 1 gallon of X and £2 to produce 1 gallon of Y. Given that the total cost of production is £C,

b express C in terms of x and y.

The company wishes to minimise the total cost.

c Using a graphical method, solve the resulting linear programming problem. Find the optimal values of x and y and the resulting total cost.

d Find the maximum cost of production for all possible choices of x and y which satisfy the inequalities you wrote down in part **a**.

(E)

26 *

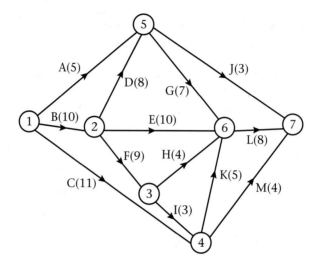

The diagram shows the activity network used to model a building project. The activities are represented by the edges. The number in brackets on each edge represents the time, in days, to complete the activity.

a Calculate the early time and the late time for each event.

b Calculate the total float for each activity.

c Hence determine the critical activities. Write down the length of the critical path.

Owing to the breakdown of a piece of equipment the time taken to complete activity H increases to 9 days.

d Obtain the new critical path and its length.

(E)

Examination style paper

Photocopy masters are available for questions marked *.

1 * a Define the term 'alternating path' (2)

Six health and safety inspectors, A, B, C, D, E and F must visit six places of work, 1, 2, 3, 4, 5, and 6. The table shows the possible allocations of inspectors to places of work.

Inspector	Places of work
A	3 or 4
B	1, 2 or 5
C	2 or 3
D	1 or 6
E	4
F	6

b Draw a bipartite graph to model this situation (1)

Initially A is assigned to 4, B is assigned to 2, C is assigned to 3 and D is assigned to 6.

c Draw a bipartite graph to illustrate this initial matching. (1)

d Starting from this initial matching, find an alternating path, starting from E, to form an improved matching. List your improved matching. (3)

e Starting from your improved matching apply the maximum matching algorithm once again to find a complete matching. (3)

(Total 10 marks)

2 Hettie, Leo, Ramin, Amro, Tom, Jing, Yonnie, Sue, Mark

a Use a bubble sort to produce this list of names in alphabetical order. You must give the state of the list after each pass. (5)

b Use the binary search algorithm to locate the name Ramin. (4)

(Total 9 marks)

3   8   14   5   11   10   3   6   12

The numbers in the list represent the lengths, in metres, of eight lengths of cable. The lengths are to be cut from rolls which each hold 20 m of cable.

a Obtain a lower bound for the number of rolls needed to supply the eight lengths of cable. (2)

b Use the first-fit bin packing algorithm to determine which lengths should be cut from each roll. (3)

c Use the first-fit decreasing bin packing algorithm to determine which lengths should be cut from each roll. (3)

(Total 8 marks)

4*

	A	B	C	D	E	F
A	–	54	22	40	36	36
B	54	–	30	26	16	22
C	22	30	–	20	14	14
D	40	26	20	–	18	24
E	36	16	14	18	–	6
F	36	22	14	24	6	–

The table shows the costs, in pounds, of travelling between six towns, A, B, C, D, E and F.

a Use Prim's algorithm, starting at A, to find a minimum spanning tree for this table of costs. You must list the arcs that form your tree in the order that they are selected. (3)

b Draw your tree and state its weight. (2)

(Total 5 marks)

5*

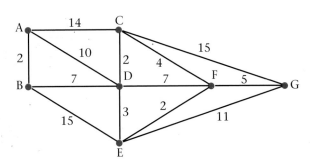

The diagram shows a network of roads. The number on each arc represents the length of that road in km.

a Use Dijkstra's algorithm to find the shortest path from A to G. Complete the boxes on the worksheet to show your working. State your shortest path and its length. (6)

b Explain how you determined the shortest route from your labelled diagram. (2)

c Find a shortest route from A to G that avoids E, and state its length. (3)

(Total 11 marks)

6

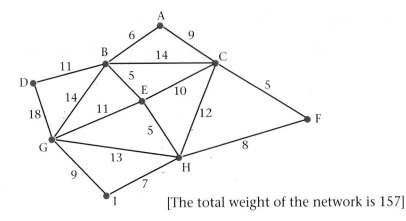

[The total weight of the network is 157]

The diagram on the previous page shows a network of pipes. The number on each arc represents the length of that pipe in metres.

The network must be inspected for leaks.

The inspection route must start and finish at A and the length of the route must be minimised.

Each pipe must be traversed at least once.

a Use an appropriate algorithm to find the length of the shortest inspection route. You should make your method and working clear. (5)

The inspector lives close to B and decides to start the inspection from there. He may finish at any vertex.

b State the vertex at which his route should finish in order to minimise the length of his route. Justify your answer. (3)

(Total 8 marks)

7 *

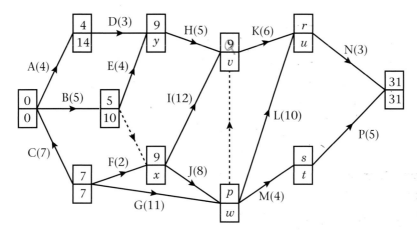

The network in the diagram above shows the activities that need to be undertaken to complete a project. Each activity is represented by an arc. The number in brackets is the duration of the activity in days. The early and late event times are shown at each vertex.

a Complete the diagram on the worksheet to show the values of P, Q, R, S, T, U, V, W, X and Y. (4)

b List the critical activities. (1)

c Calculate the total float on activities E and H. You must show the numbers you used in your calculation. (3)

d Draw a cascade (Gantt) chart for the project. (5)

e Write down the activities that *must* be happening at midday on day 13. (2)

(Total 15 marks)

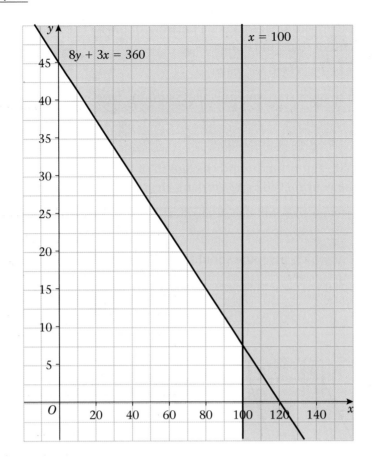

8* The diagram shows two of the constraints in a linear programming problem.

$$8y + 3x \leqslant 360$$

and

$$x \leqslant 100$$

two further constraints are

$$y \leqslant 25$$

and

$$4y + x \leqslant 140$$

a Add two lines and shading to the copy of the diagram on the worksheet to represent these inequalities. (3)

b Hence determine and label the feasible region, R. (1)

For this linear programming problem the objective function is

$$F = 3y + x$$

c Add an objective line to your graph and use it to determine the optimal number of variables x and y. (4)

d State the optimal value of the objective. (1)

(Total 9 marks)

Answers

Exercise 1A

1 a 1 $\frac{a}{b} = \frac{9}{4}$ $\frac{c}{d} = \frac{4}{3}$ $a = 9, b = 4, c = 4, d = 3$
    2 $e = ad = 9 \times 3 = 27$
    3 $f = bc = 4 \times 4 = 16$
    4 answer is $\frac{27}{16}$
 b It divides the first fraction by the second fraction.
2 a 1 $A = 1$ $n = 1$
    2 Print 1
    3 $A = 1 + 2 \times 1 + 1 = 4$
    4 $n = 2$
    5 $2 \leqslant 10$ go to 2
    2 Print 4
    3 $A = 4 + 2 \times 2 + 1 = 9$
    4 $n = 3$
    5 $3 \leqslant 10$ go to 2
    2 Print 9
    3 $A = 9 + 2 \times 3 + 1 = 16$
    4 $n = 4$
    5 $4 \leqslant 10$ go to 2
    2 Print 16
    3 $A = 16 + 2 \times 4 + 1 = 25$
    4 $n = 5$
    5 $5 \leqslant 10$ go to 2
    2 Print 25
    3 $A = 25 + 2 \times 5 + 1 = 36$
    4 $n = 6$
    5 $6 \leqslant 10$ go to 2
    2 Print 36
    3 $A = 36 + 2 \times 6 + 1 = 49$
    4 $n = 7$
    5 $7 \leqslant 10$ go to 2
    2 Print 49
    3 $A = 49 + 2 \times 7 + 1 = 64$
    4 $n = 8$
    5 $8 \leqslant 10$ go to 2
    2 Print 64
    3 $A = 64 + 2 \times 8 + 1 = 81$
    4 $n = 9$
    5 $9 \leqslant 10$ go to 2
    2 Print 81
    3 $A = 81 + 2 \times 9 + 1 = 100$
    4 $n = 10$
    5 $10 \leqslant 10$ go to 2
    2 Print 100
    3 $A = 100 + 2 \times 10 + 1 = 121$
    4 $n = 11$
    5 $11 \nleqslant 10$
    6 stop
    Output 1, 4, 9, 16, 25, 36, 49, 64, 81, 100
 b It finds the squares of the first 10 natural numbers.

3 a i

Step	A	r	C	$\lvert r - C \rvert$	s	Print r
1	253	12				
2			21.083			
3				9.083		
4					16.542	
5		16.542				
6 → 2			15.294			
3				1.248		
4					15.918	
5		15.918				
6 → 2			15.894			
3				0.024		
4					15.906	
5		15.906				
6 → 2			15.906			
3 → 7				0		
7						$r = 15.906$
8 stop						

ii

Step	A	r	C	$\lvert r - C \rvert$	s	Print r
1	79	10				
2			7.9			
3				2.1		
4					8.95	
5		8.95				
6 → 2			8.827			
3				0.123		
4					8.889	
5		8.889				
6 → 2			8.887			
3 → 7				0.002		
7						Print 8.889

iii

Step	A	r	C	$\lvert r - C \rvert$	s	Print r
1	4275	50				
2			85.5			
3				35.5		
4					67.75	
5		67.75				
6 → 2			63.100			
3				4.65		
4					65.425	
5		65.425				
6 → 2			65.342			
3				0.083		
4					65.384	
5		65.384				
6 → 2			65.383			
3 → 7				0.01		
7						Print 65.384

b Finds the square root of A.

4 a

A	B
244	125
122	250
61	500
30	1000
15	2000
7	4000
3	8000
1	16 000
Total	30 500

b

A	B
125	244
62	488
31	976
15	1952
7	3904
3	7808
1	15 616
Total	30 500

c

A	B
256	123
128	246
64	492
32	984
16	1968
8	3936
4	7872
2	15 744
1	31 488
Total	31 488

Exercise 1B

1 a

a	b	c	d	d < 0?	d = 0	x
4	−12	9	0	No	Yes	1.5

Equal roots $x = 1.5$

b

a	b	c	d	d < 0?	d = 0?	x_1	x_2
−6	13	5	289	No	No	$-\frac{1}{3}$	$\frac{5}{2}$

Roots are $-\frac{1}{3}$ and $\frac{5}{2}$

c

a	b	c	d	d < 0
3	−8	11	−68	Yes

No real roots

2 a i output is 21   **ii** output is 5
 b It will find the largest number in the list.
 c box 6 – changed to 'Is $n < 8$?'
3 a i print 13   **b** Finds the HCF.
 ii print 17
 iii print 15
4 a 1.8041   **b** 1.8041 – same root

Exercise 1C

1 a 2 3 4 5 6 7 8
 b 8 7 6 5 4 3 2
2 a 7 11 14 17 18 20 22 25 29 30
 b 30 29 25 22 20 18 17 14 11 7
3 c C E H J K L M N P R S
 b C E H J K L M N P R S
4 a

Amy	93
Greg	91
Janelle	89
Sophie } Dom	77
Lucy	57
Alison	56
Annie	51
Harry	49
Josh	37
Alex	33
Sam	29
Myles	19
Hugo	9

 b 93 91 89 77 77 57 56 51 49 37 33 29 19 9
5 The quick sort should be notably faster than the bubble sort.

Exercise 1D

1 a Connock at position 2.
   b Walkey at position 6.
   c Peabody not in list.
2 a 21 at position 11.
   b 5 not in list.
3 a Fredco at position 7.
   b Matt at position 15.
   c Elliot not in list.
4 a P at position 16.
   b At most 5 passes.
5 a 7
   b 10
   c 14

Exercise 1E

1 a 5 bins
   b i   Bin 1: 18 + 4 + 23 + 3
           Bin 2: 8 + 27
           Bin 3: 19 + 26
           Bin 4: 30
           Bin 5: 35
           Bin 6: 32
      ii Bin 1: 35 + 8 + 4 + 3
           Bin 2: 32 + 18
           Bin 3: 30 + 19
           Bin 4: 27 + 23
           Bin 5: 26
      iii Bin 1: 32 + 18
           Bin 2: 27 + 23   } full bins
           Bin 3: 35 + 8 + 4 + 3
           Bin 4: 19 + 26
           Bin 5: 30
2 a Bin 1: A(30) + B(30) + C(30) + D(45) + E(45)
       Bin 2: F(60) + G(60) + H(60)
       Bin 3: I(60) + J(75)
       Bin 4: K(90)
       Bin 5: L(120)
       Bin 6: M(120)
   b Bin 1: M(120) + I(60)
       Bin 2: L(120) + H(60)
       Bin 3: K(90) + J(75)
       Bin 4: G(60) + F(60) + E(45)
       Bin 5: D(45) + C(30) + B(30) + A(30)
   c Lower bound 5 so b optimal.
   d Bin 1: M(120)
       Bin 2: L(120)
       Bin 3: A(30) + K(90)
       Bin 4: F(60) + G(60)   } full bins
       Bin 5: H(60) + I(60)
       Bin 6: J(75) + E(45)
       Bin 7: B(30) + C(30) + D(45)
3 a Not possible in 3 bins.
   b Not possible in 3 bins.
   c Bin 1: H(14) + G(12) + A(4)
       Bin 2: C(13) + E(13) + F(4)
       Bin 3: J(11) + B(7) + D(6) + I(6)
4 a Bin 1: A(3) + B(3) + C(4) + D(4)
       Bin 2: E(4) + F(4) + G(5)
       Bin 3: H(5) + I(5)
       Bin 4: J(7) + K(8)
       Bin 5: L(8)
       5 rolls used and 15 m wasted.

   b Bin 1: L(8) + J(7)
       Bin 2: K(8) + I(5)
       Bin 3: H(5) + G(5) + F(4)
       Bin 4: E(4) + D(4) + C(4) + B(3)
       Bin 5: A(3)
       5 rolls used and 15 m wasted.
   c Bin 1: L(8) + J(7)
       Bin 2: G(5) + H(5) + I(5)
       Bin 3: C(4) + D(4) + E(4) + B(3)
       Bin 4: K(8) + F(4) + A(3)
       4 rolls used and no wastage.
5 a Bin 1: H(25) + A(8)
       Bin 2: G(25)
       Bin 3: F(24) + B(16)
       Bin 4: E(22) + C(17)
       Bin 5: D(21)
   b Lower bound is 4.
   c There are 5 programs over 20 MB. It is not possible
       for any two of these to share a bin. So at least 5 bins
       will be needed, so 4 will be insuficient.

Exercise 1F

1 1 2 15 16 27 38
2 a 42 41 31 26 25 22   b 15
3 2 4 8 9 13 15 17
4 a 115 111 103 98 93 82 81 77 68
   b i   Bin 1: 115 + 82
           Bin 2: 111 + 81
           Bin 3: 103 + 93
           Bin 4: 98 + 77
           Bin 5: 68
      ii No room in bin 1 (3 left) or bin 2 (8 left) or
           bin 3 (4 left) but room in bin 4.
5 a Rank the times in descending order and use them
       in this order. Number the bins, starting with bin 1
       each time.
       Bin 1: 100
       Bin 2: 92
       Bin 3: 84 + 30
       Bin 4: 75 + 42
       Bin 5: 60 + 52
       unused tape: 65 minutes
   b There is room on tape 2 for one of the 25-minute
       programmes but no room on any tape for the
       second programme.
6 a The two 1.2 m lengths cannot be 'made up' to
       2 m bins since there are only 3 × 0.4 m lengths.
       Two of these can be used to make a full turn,
       1.2 + 0.4 + 0.4, but the second 1.2 m cannot be
       made up to 2 m since there is only one remaining
       0.4 m length.
   b Bin 1: 1.6 + 0.6
       Bin 2: 1.4 + 1
       Bin 3: 1.2 + 1.2
       Bin 4: 1 + 1 + 0.4
       Bin 5: 0.6 + 0.6 + 0.6 + 0.4
       Bin 6: 0.4
   c Bin 1: 1.6 + 0.4 + 0.4
       Bin 2: 1.4 + 1
       Bin 3: 1.2 + 1.2
       Bin 4: 1 + 1 + 0.4
       Bin 5: 0.6 + 0.6 + 0.6 + 0.6
7 a output 4.8
   b It selects the number nearest to 5.
   c It would select the number furthest from 5.

Exercise 2A

1 a For example

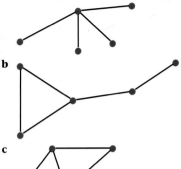

b

c

2 a is not simple. There are two edges connecting C with D.
b and **c** are simple.
d is not simple. There is a loop attached to U.

3 a and **c** are connected.
b is not connected, there is no path from J to G, for example.
d is not connected, there is no path from W to Z, for example.

4 a Any four of these

F A B D   F E D
F A C B D   F E C B D
F A B C E D   F E C A B D
F A C E D

b Here are examples. (These all start at F, but you could start at any point.) There are others.

F A B D E F   F A B D B A F   F A C E D E C A F
F E D B A F   F E D E F   F A C E D E F
F A C B D E F   F A C B D B C A F   F E D E C A F
F E D B C A F   F E C B D B C E F   F E C B D E F
F E D B C E F

c

Vertex	A	B	C	D	E	F
Degree	3	3	3	2	3	2

d Here are examples. (There are many others.)

i A

ii F

B

E

C

iii A —— B

C   D

e Sum of degrees = 3 + 3 + 3 + 2 + 3 + 2 = 16
number of edges = 8
sum of degrees = 2 × number of edges for this graph

5 a

Vertex	J	K	L	M	N	P	Q	R	S
Degree	1	2	3	1	1	4	2	1	1

Here are some possible subgraphs (there are many others).

i L ———————— P

S

ii   Q

P   S

iii   L

J   S

Sum of degrees = 1 + 2 + 3 + 1 + 1 + 4 + 2 + 1 + 1 = 16
number of edges = 8
sum of degrees = 2 × number of edges for this graph

b This graph is a tree so there will only be one path between any two vertices.

6 a For example

6 vertices all even
5 of degree 2
1 of degree 4

b For example

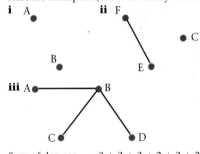

3 vertices all even, all of order 2
The sum sum of degrees = 2 × number of edges, so the sum of degrees must be even. Any vertices of odd degree must therefore 'pair up'. So there must be an even number of vertices of odd degree.

7

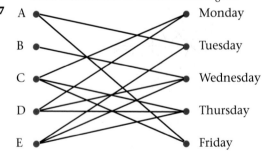

A   Monday
B   Tuesday
C   Wednesday
D   Thursday
E   Friday

8

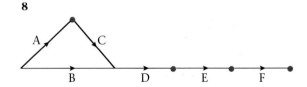

Exercise 2B

1 **a** and **b** are trees.

b is not a tree, it is not a connected graph.

d is not a tree, it contains a cycle.

2 i
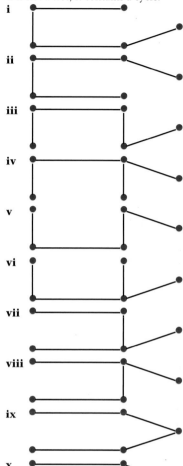

ii

iii

iv

v

vi

vii

viii

ix

x

xi

3

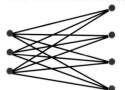

k_8

Each vertex will have a degree of $(n-1)$.

4

$k_{3, 4}$

There will be nm edges.

5 **a**, **e** and **h** are isomorphic.

b and **i** are isomorphic.

c and **g** are isomorphic.

d and **f** are isomorphic.

6

	A	B	C	D	E	F	G
A	0	1	1	0	2	0	0
B	1	2	1	0	0	0	0
C	1	1	0	2	0	1	0
D	0	0	2	0	0	1	1
E	2	0	0	0	0	1	0
F	0	0	1	1	1	0	1
G	0	0	0	1	0	1	0

7 **a**

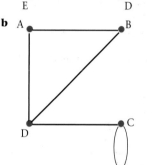

b

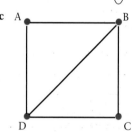

c

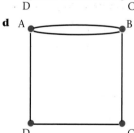

d

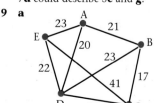

8 7**c** could describe 5**a**, **e** and **h**.

7**d** could describe 5**c** and **g**.

9 **a**

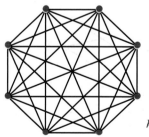

b

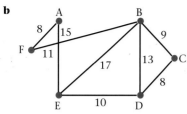

10

	A	B	C	D	E	F
A	—	14	11	—	—	—
B	—	—	10	13	11	—
C	—	—	—	12	—	—
D	—	—	—	—	—	10
E	—	—	—	—	—	7
F	—	—	—	—	9	—

Exercise 2C

1

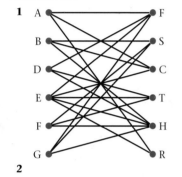

2

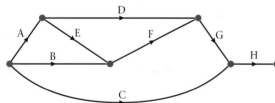

3 a

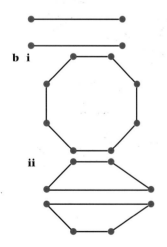

b i

ii

iii

4

	A	B	C	D	E	F
A	—	20	18	16	—	—
B	20	—	15	—	—	50
C	18	15	—	10	20	30
D	16	—	10	—	23	—
E	—	—	10	23	—	25
F	—	50	30	—	25	—

5 Here are examples. (There are many other solutions.)

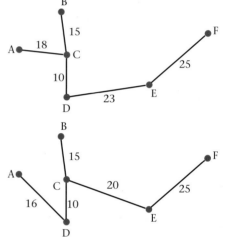

6 $E = V - 1$ (or $V = E + 1$)

Exercise 3A

1 a

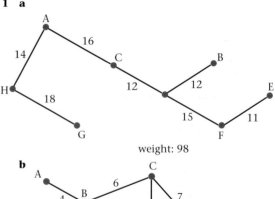

weight: 98

b

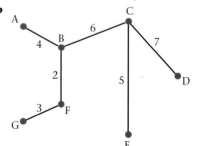

weight: 27

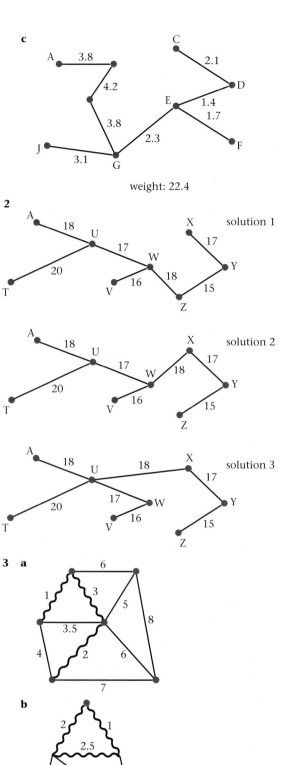

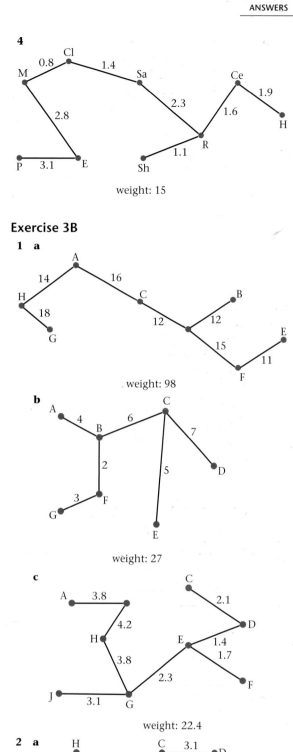

c

weight: 22.4

2

solution 1

solution 2

solution 3

3 a

b

4

weight: 15

Exercise 3B

1 a

weight: 98

b

weight: 27

c

weight: 22.4

2 a

b £17 340

3

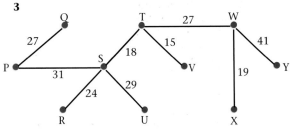

b £231 000

4 Solution 1

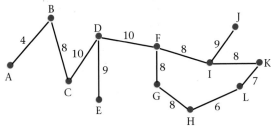

Solution 2

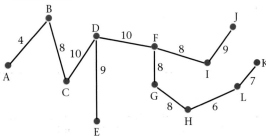

Solution 3

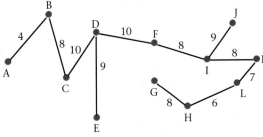

Solution 4

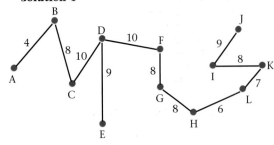

Exercise 3C

1 a Arcs in order
AF (9)
FB (14)
AC (20)
AE (25)
DE (26)
weight = 94

b Arcs in order
RS (28)
ST (16)
SU (19)
UV (37)
weight = 100

2 Arcs in order
BS (49)
SM (44)
SN (56)
NL (37)
weight = 186

3 a cost = €1014

b

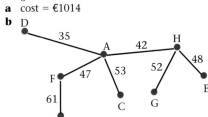

c i It is cheaper to translate from E to H then from H to G at a cost of 48 + 52 = 100 euro rather than 159 euro per 1000 words.

ii A direct translation is likely to be more accurate then a translation via another language.

4 a order of arcs
XE (26)
EG (18)
EH (23)
HA (25)
AF (20)
FB (16)
AD (22)
FC (24)
CI (26)

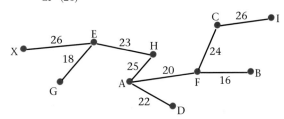

b order of arcs
XE (26)
EG (18)
EH (23)
HF (31)
FB (16)
FC (24)
BD (24)
CI (26)

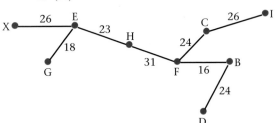

Exercise 3D

1 a Shortest route: S – B – E – H – G – C – F – I
Length of shortest route: 20
b Shortest route: S – A – B – D – F – H – T
Length of shortest route: 15
c Shortest route: S – C – E – H – T
Length of shortest route: 31
2 a A to Q   A – F – E – I – N – P – Q   Length 37
b A to L   A – F – G – M – L   Length 19
c M to A   M – G – F – A   Length 17
d P to A   P – N – I – E – F – A   Length 24
3 Shortest route: A – B – E – C – G – F   Length 38
4 Shortest route: S_1 – B – C – F – E – G – T
Length of shortest route: 1660
5 a 94 – 18 = 76 EH   76 – 24 = 52 CE
52 – 22 = 30 BC   30 – 30 = 0 AB
Shortest route A to H: A – B – C – E – H
Length 94
b Shortest route A to H via G: A – D – G – H
Length 96
c Shortest route A to H not using C E: A – D – F – E – H
Length 95
6 Shortest route: S – B – C – T
Length of shortest route: 10

Exercise 3E

1 a i Arcs are labelled with initial letters of the nodes.
CK add to tree
SH add to tree
CE add to tree
EK reject
CH add to tree
HW add to tree
CS reject
HQ add to tree
QS reject
QD add to tree
KS reject
EW reject
ii EC
CK
CH
HS
HW
HQ
QD
b

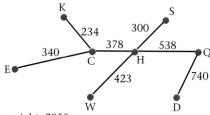

weight: 2953

2 a i LT
MT
MQ
NQ
ST
QR
NP

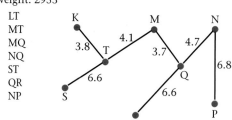

ii MQ (3.7) add to tree
LT (3.8) add to tree
MT (4.1) add to tree
NQ (4.7) add to tree
MN (5.3) reject
ST (6.6) add to tree
QR (6.6) add to tree
NP (6.8) add to tree
reject remaining arcs
b Start off the tree with QT and PR then apply Kruskal's algorithm. Prim's algorithm requires the 'growing' tree to be connected at all times. When using Kruskal's algorithm the tree can be built from non-connected sub-trees.

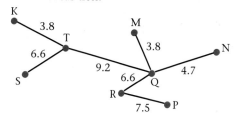

3

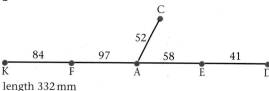

length 332 mm

4 Arcs in order: Entrance 2–office; Entrance 2–Entrance 4; Entrance 4–Entrance 3; Office–Entrance 1

length: 3112 m

5 i EF (8)   add to tree
CD (9)   add to tree
EJ (12) add to tree
FJ (14) reject
GI (14) add to tree
CG (17) add to tree
DG (18) reject
BC (19) add to tree
FI (20) add to tree
HK (22) add to tree
EK (24) add to tree
CF (24) reject
JK (26) reject
AE (27) add to tree
AB (29) } reject remaining arcs
AH (30) }
ii

iii weight: 172
b $V = E + 1$

6 a

weight = 45 so 4500 m needed

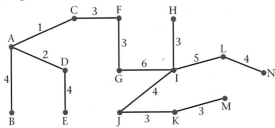

b weight = 47 so 4700 m

7 a i Possible paths are A – H – G – E – I – K
  and A – H – J – I– K } anyone accepted
  and A – B – C – K

ii
44 − 9 = 35	IK		44 − 9 = 35	IK
35 − 10 = 25	EI		35 − 17 = 18	JI
25 − 10 = 15	GE	*or*	18 − 8 = 10	HJ
15 − 5 = 10	HG		10 − 10 = 0	AH
10 − 10 = 0	AH			

or

44 − 12 = 32	CK
32 − 25 = 7	BC
7 − 7 = 0	AB

b A – H – G – E – I – K and A – H – J – I – K
and A – B – C – K

c The arcs could be roads.
The nodes could be junctions.
The number on each arc could be distance in km.
The network, together with Dijkstra's algorithm,
could be used to find the shortest route from A to K.

8 Order of vertex labelling:
S C B A E D G F T
Route: S – C – E – F – T
411 − 101 = 310   FT
310 − 123 = 187   EF
187 − 85 = 102   CE
102 − 102 = 0   SC

Exercise 4A

1 a
vertex	A	B	C	D	E	F
valency	2	3	2	3	3	3

There are 4 nodes with odd valency so the graph is
neither Eulerian nor Semi-Eulerian.

b
vertex	G	H	I	J	K
valency	3	4	3	2	4

There are precisely 2 nodes of odd degree (G and I)
so the graph is *semi-Eulerian.*
A possible route starting at G and finishing at I is:
G – H – K – I – J – K – G – H – I.

c
vertex	L	M	N	P	Q	R
valency	2	4	2	4	2	4

All vertices have even valency, so the graph is
Eulerian.
A possible route starting and finishing at L is:
L – M – N – P – M – R – P – Q – R – L.

2 a i
vertex	A	B	C	D	E	F	G	H
valency	4	2	4	2	2	4	2	2

ii
vertex	A	B	C	D	E	F	G
valency	4	4	2	4	2	4	4

b i A possible route is:
  A – B – C – A – F – C – E – G – H – F – D – A.
ii A possible route is:
  A – C – F – A – B – E – G – B – D – G – F – D – A.

3 a i
vertex	R	S	T	U	V	W
valency	2	2	3	3	2	2

Precisely 2 vertices of odd valency (T and U) so
semi-Eulerian.

ii
vertex	H	I	J	K	L	M	N
valency	2	4	3	2	3	4	4

Precisely two nodes of odd degree (J and L) so semi-
Eulerian.

b i A possible route starting of T and finishing at U is:
  T – R – S – U – W – V – T – U.
ii A possible route starts at J and finishes at L:
  J – K – L – M – J – I – M – N – I – H – N – L.

4 a
vertex	A	B	C	D	E	F	G	H	I
valency	1	3	1	1	4	1	1	3	1

There are more than 2 vertices of odd degree so the
graph is not traversable.

b
vertex	U	V	W	X	Y	Z
valency	3	3	2	5	2	3

There are more than 2 nodes of odd degree so the
graph is not traversable.

5 In each case there are either zero, or an even number
of, vertices with odd valency.

6 If a graph is traversable we will approach each vertex
on one edge and must leave on a different one.

'in' edge

'out' edge

a This means that the edges must be in pairs at each
vertex, an 'out' edge for each 'in' edge, and since
the graph is traversable there will be no edges left
over. So the valency of each vertex will be even if
we return to the start.

b For routes that start and finish at different vertices
there will be an unpaired 'out' edge from the start
vertex which will be balanced by an unpaired 'in'
edge at the finish vertex. So these two vertices will
have odd valency, but all others will be even.

7 a

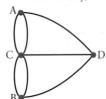

b
vertex	A	B	C	D
valency	3	3	5	3

There are more than two odd nodes, so the graph is
not traversable.

c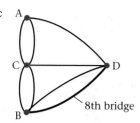

8th bridge

We will start of A and finish at C so these still need to have odd valency. We can only have two odd valencies so B and D must have even valencies (see table).

We need to change the valency of B and of D. So we build a bridge from B to D.

vertex	A	B	C	D
valency with 7 bridges	odd	*odd*	odd	*odd*
valency wanted	odd	*even*	odd	*even*

d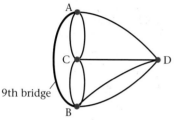

9th bridge

We will start at B and finish at C so these vertices need to be the two vertices with odd valency. We need A and D to have even valency (see table).
We need to change the valency of node A and of node B.
So we build a bridge from A to B.

vertex	A	B	C	D
valency with 8 bridges	*odd*	even	odd	even
valency wanted	*even*	odd	odd	even

e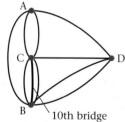

10th bridge

All vertices now need to have even valency.
This means we need to change the valencies of nodes B and C.
So the 10th bridge needs to be built from B to C.

vertex	A	B	C	D
valency with 9 bridges	even	*odd*	*odd*	even
valency wanted	even	*even*	*even*	even

Exercise 4B

1 a All valencies are even, so the network is traversable and can return to its start.

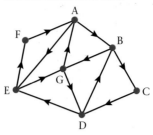

A possible route is:
A − B − C − D − B − G − D − E − G − A − E − F − A.
length of route = weight of network
= 285

b The valencies of D and E are odd, the rest are even. We must repeat the shortest path between D and E, which is the direct path DE.
We add this extra arc to the diagram.

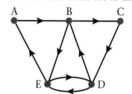

A possible route is:
A − B − C − D − E − D − B − E − A.
length of route = weight of network + arc DE
= 61 + 11
= 72

c The degrees of B and E are odd, the rest are even. We must repeat the shortest path from B to E. By inspection this is BCE, length 260.
We add these extra arcs to the diagram.

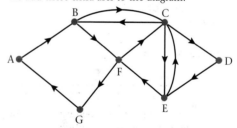

A possible route is:
A − B − C − D − E − C − B − F − C − E − F − G − A.
length of route = weight of network + BCE
= 1005 + 260
= 1265

d The order of B and G are odd, the rest are even. We must repeat the shortest path from B to G. By inspection this is BDHG, length 183.
We add these arcs to the diagram.

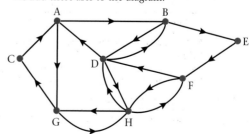

A possible route is:
A − B − E − F − D − B − D − A − G − H − F − H −
D − H − G − C − A.
length of route = weight of network + BDHG
= 995 + 183
= 1178

2 a Odd valencies at B, D, E and F.
Considering all possible pairings and their weights.
BD + EF = 130 + 85 = 215 ← least weight
BE + DF = 110 + 178 = 288
BF + DE = 125 + 93 = 218

> Shortest route
> D to F, is DEF.

We need to repeat arcs BD and EF.
The length of the shortest route
= weight of network + 215
= 908 + 215
= 1123
Adding BD and EF to the diagram gives.

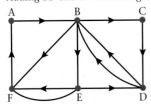

A possible route is:
A − B − C − D − B − E − D − B − F − E − F − A.

b Odd valencies at C, D, E and G
Considering all possible pairings and their weights
CD + EG = 130 + 75 = 205
CE + DG = 157 + 92 = 249
CG + DE = 82 + 120

> Shortest route
> from C to E is
> CGE.

= 202 ← least weight
We need to repeat arcs CG and DE.
The length of the shortest route
= weight of network + 202
= 938 + 202
= 1123
Adding CG and DE to the diagram gives.

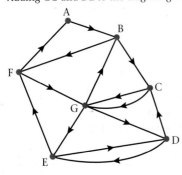

A possible route is:
A − B − C − G − D − C − G − E − D − E − F − G −
B − F − A.

c Odd degrees at B, D, G and I.
Considering all possible pairings and their weight
BD + GI = 22 + 30 = 52
BG + DI = 37 + 42 = 79
BI + DG = 32 + 18

> Shortest route
> BG = BEG
> DI = DEI

= 50 ← least weight
We need to repeat arcs BI and DG.
The length of the shortest route
= weight of network + 50
= 326 + 50
= 376

Adding BI and DG to the diagram gives.

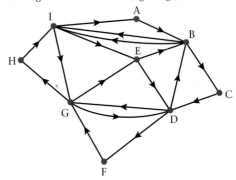

A possible route is
A − B − C − D − B − I − E − B − I − G − D − F − G
− E − D − G − H − I − A.

3 a Odd degrees are B, D, E and F.
Considering all possible pairings and their weight
BD + EF = 250 + 200 + = 450 ← least weight
BE + DF = 350 + 380 = 730
BF + DE = 300 + 180 = 480
We need to repeat arcs BD and EF.
Adding these to the diagram gives

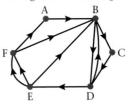

A possible route is:
A − B − C − D − B − D − E − F − B − E − F − A.
length = 1940 + 450 = 2360.

b We will still have two odd valencies.
We need to select the pair that gives the least path.
From part **a** our six choices are
BD (250), EF (200), BE (350), DF (380), BF (300) and
DE (180).
The shortest is DE (180) so we choose to repeat this.
It is the other two vertices (B and F) that will be our
start and finish.
For example, start at B, finish at F
length of route = 1910 + 180 = 2090

4 a Each arc must be traversed twice, whereas in the
standard problem each arc need only be visited
once.
This has the same effect as doubling up all the
edges
The length of the route = 2 × weight of network
= 2 × 89 = 178 km

b Odd nodes C, D, E, G.
Considering all possible pairings.
CD + EG = 15 + 5 = 20
CE + DG = 15 + 3 = 18
CG + DE = 10 + 7 = 17 ← least weight

> Shortest routes
> CD is CGD
> CE is CGE

We need to repeat arcs CG and DE.

Adding these to the network

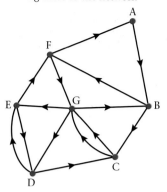

A possible route is
A − B − C − G − D − C − G − E − D − E − F − G − B − F − A.

c If EG is omitted E and G become even and the only odd valencies are at C and D.
We must repeat the shortest path between C and D, CGD.
The new length = 89 + 13 = 102 km

Mixed Exercise 4C

1 a

vertex	A	B	C	D	E	F	G	H	I	J
degree	2	2	2	5	4	3	2	3	3	4

b DF + HI = 19 + 36 = 55
DH + FI = 22 + 27 = 49 ← least weight
DI + HF = 46 + 41 = 87

> Shortest routes: D to I is DFI; H to F is HDF

Repeat DH and FI
Add these to the network to get

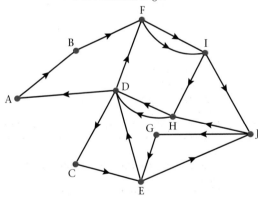

A possible route is
A − B − F − I − J − G − E − J − H − D − F − I − H − D − C − E − D − A.

c length = 725 + 49 = 774

2 Odd vertices Q, R, T, V
Considering all possible pairings and their weight
QR + TV = 104 + 189 = 293
QT + RV = 153 + 115 = 268 ← least weight
QV + RT = 163 + 123 = 286

> Shortest route from Q to T is QST.

Repeat arcs QS, ST and RV.

Add these to the network

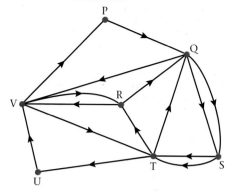

A possible route is
P − Q − S − T − Q − S − T − R − Q − V − R − V − T − U − V − P.
length of route = 1890 + 268
= 2158

3 a

vertex	A	B	C	D	E	F	G	Odd valencies
degree	4	3	3	3	4	4	3	at B, C, D and G.

b Considering all possible pairings and their weight
BC + DG = 7 + 16 = 23
BD + CG = 20 + 18 = 38
BG + CD = 16 + 13 = 29
BC, DF and FG should be repeated.

> Shortest routes: OG is DFG; BD is BCD; CG is CEFG; BG is BAG.

Adding these arcs to the network gives

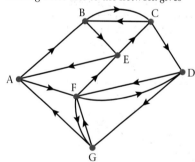

A possible route is
A − B − C − B − E − C − D − F − D − G − F − G − A − F − E − A.

c Length = 108 + 23 = 131

4 a

vertex	A	B	C	D	E	F	G	H	I
valency	2	3	4	3	4	2	6	3	3

Odd valencies at B, D, H and I.

b Considering all possible complete pairings and their weight
BD + HI = 7.2 + 3.4 = 10.6
BH + DI = 7.6 + 4 = 11.6
BI + DH = 5.6 + 4.3 = 9.9 ← least weight

> Shortest routes: BD is BED; BH is BEGH, DI is DGI; BI is BEI, DH is DGH.

Repeat BE, EI and DG, GH.

Adding these arcs to the network gives

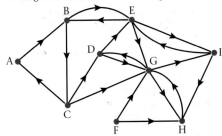

A possible route is:
A − B − E − I − H − G − I − E − B − C − D − G −
D − E − G − H − F − G − C − A.

c length = 51.4 + 9.9 = 61.3 km

d If BD is included B and D now have even valency.
Only H and I have odd valency.
So the shortest path from H to I needs to be
repeated.
Length of new route = 51.4 + BD + path from H to I
= 51.4 + 6.4 + 3.4
= 61.2 km
This is (slightly) shorter than the previous route so
choose to grit BD since it saves 0.1 km.

5 a The route inspection algorithm
(method as shown in main text page 69)

b Odd valencies B, C, E, H.
Considering all possible complete pairings and their
weight
BC + EH = 68 + 150 = 218
BE + CH = 95 + 73 = 168 ← least weight
BH + CE = 141 + 85 = 226

Shortest routes: BE is BDE; BH is BCH, CE is CDE

Repeat BD, DE and CH
Adding these arcs to the network gives

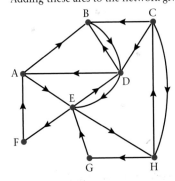

A possible route is:
A − B − D − B − C − H − C − D − E − D − A − E −
H − G − E − F − A.
length = 1011 + 168
= 1179 m

c This would make B the start and C the finish.
We would have to repeat the shortest path between
E and H only.
New route = 1011 + 150 = 1161 m.
1161 < 1179
So this would decrease the total distance by 18 m.

6 a The route inspection algorithm – description in
main text on page 69.

b Odd vertices B, D, F, H
Considering all complete pairings
BD + FH = 14 + 15 = 29
BF + DH = 10 + 26 = 36
BH + DF = 12 + 16 = 28 ← least weight

The shortest route DH is DBH.

Repeat BH and DF.
Adding these arcs to the network gives

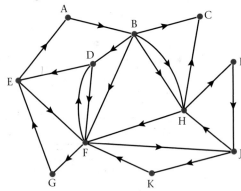

A possible route is:
A − B − H − C − B − H − I − J − H − F − J − K − F
− B − D − F − D − E − F − G − E − A.

c length of route = 249 + 28 = 277

d i We will still have to repeat the shortest path
between a pair of the odd nodes.
We will choose the pair that requires the
shortest path.
The shortest path of the six is BF (10).
We will use D and H as the start and finish
nodes.

ii 259

e Each edge, having two ends, contributes two to the
sum of valencies for the network.
Therefore the sum = 2 × number of edges
The sum is even so any odd valencies must occur
in pairs.

Review exercise 1

1 For example,

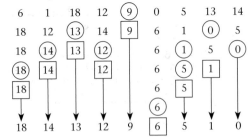

Datchet (18), Wraysbury (14), Staines (13) Feltham
(12), Halliford (9) Ashford (6), Poyle (5), Colnbrooke
(1), Laleham (0)

2 a All arcs are to be traversed twice, this is, in effect,
repeating each arc. So all valencies are even

b E.g. A − B − D − G − F − G − D − C − E − A − E −
C − A − F − E − F − B − F − A − B − D − C − A (all
correct routes will have 23 letters in their name)
length = 2 × 6 = 12 km.

3 a 1st pivot is $\left\lceil \dfrac{1+10}{2} \right\rceil \to 6$ FEW, SABINE after FEW

Rejecting: $1 - 6$ list is now
7 OSBORNE
8 PAUL
9 SWIFT
10 TURNER

2nd pivot is $\left\lceil \dfrac{7+10}{2} \right\rceil \to 9$ SWIFT, SABINE before SWIFT.

Rejecting 9 and 10 list is now
7 OSBORNE
8 PAUL

3rd Pivot is $\left\lceil \dfrac{7+8}{2} \right\rceil \to 8$ PAUL, SABINE is after PAUL

rejecting 7 and 8 list is now empty
So SABINE is not in the list

b Either:
maximum length of list remaining is
1000, 500, 250, 125, 63, 31, 15, 7, 3, 1
so 10 iterations.
or we seek the smallest integer value of n such that
$$2^n > 1000$$
$$\log 2^n > \log 1000$$
$$n \log 2 > \log 1000$$
$$n > \frac{\log 1000}{\log 2}$$
$$n > 9.966$$
$\therefore$ n is 10

4 a

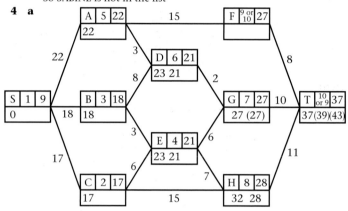

Shortest time: 37 minutes

b either $S - A - D - G - T$ or $S - B - E - G - T$
The route is not unique; there are two of them
$(S - A - D - G - T$ and $S - B - E - G - T)$

c $S - C - E - G - T$ length 39 minutes.

5 a i Prim
- Select any vertex to start the tree.
- Select the shortest arc that joins a vertex already in the tree to a vertex not yet in the tree. Add it to the tree.
- Continue selecting shortest arcs until all vertices are is the tree.

ii Kruskal
- Sort all arcs into ascending order of weight.
- Select the arc at least weight to start the tree.
- Take arcs in order of weight.
  - Reject the arc if it would form a cycle.
  - Add it to the tree if it does not form a cycle.
- Continue taking each arc in turn until all vertices are connected.

b i GH, GI, HF, FD, DA, AB, AC, DE

ii GH, AB, AC, AD, reject BD, DF, GI, reject BC, FH, reject DG, DE

c weight is 76

d Prim's algorithm
- It is easily converted into matrix form.
- It is difficult to use Kruskal's algorithm because it is difficult to check for cycles in a large network.

6 a GC, FD, GF, reject CD, ED, reject EF, BC, AG

b

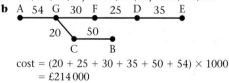

cost $= (20 + 25 + 30 + 35 + 50 + 54) \times 1000$
$= £214\,000$

7 a i

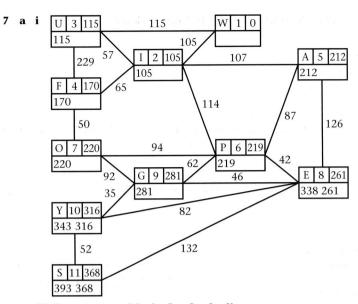

ii

$$368 - 52 = 316 \quad \text{SY}$$
$$316 - 35 = 281 \quad \text{GS}$$
$$281 - 62 = 219 \quad \text{PG}$$
$$219 - 114 = 105 \quad \text{IP}$$
$$105 - 105 = 0 \quad \text{WI}$$

iii Shortest route is W − I − P − G − S − Y

iv length 368 miles

b If we started at S, the algorithm would find the least distance from S to each vertex, so in finding S to W we could have found S to A.

8 a

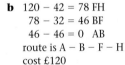

b $120 - 42 = 78$ FH

$78 - 32 = 46$ BF

$46 - 46 = 0$ AB

route is A − B − F − H

cost £120

9 a ii

ii

$43 - 10 = 33$ HJ

$33 - 18 = 15$ FH

$15 - 15 = 0$ AF

route is A − F − H − J

length 43 minutes

b Use the algorithm starting at J to find the shortest route from J to C, then add arc CA.

10 a Each edge contributes 1 to the order of the vertices at each end.

So each edge contributes 2 to the total sum of the orders.

The sum of the orders is therefore an even number.

If there were an odd number of vertices with odd order the sum would be odd. So there must always be an even (or zero) number of vertices of odd order.

b The only vertices of odd order are B and C, we have to repeat the shortest path between B and C.

If $x \geq 9$ the shortest path is BC (direct)

Weight of network + BC = 100

$(9\frac{1}{2}x - 26) + x = 100 \Rightarrow x = 12$

If $x < 9$ the shortest path is BAC of length $2x - 9$

$(9\frac{1}{2}x - 26) + 2x - 9 = 100 \Rightarrow x = 11\frac{17}{23} \nless 9$ so inconsistent.

11 a i Method:
- Start at A and use this to start the tree.
- Choose the shortest edge that connects a vertex already in the tree to a vertex not yet in the tree. Add it to the tree.
- Continue adding edges until all vertices are in the tree.   AF, FC $\left\{\begin{array}{c} \text{FB} \\ \text{or} \\ \text{BC} \end{array}\right\}$, FD, EB

ii

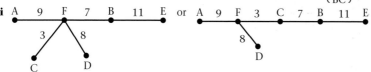

iii The tree is not unique, there are 2 of them (see above).

b i number of edges = $7 - 1 = 6$ **ii** number of vertices = $n + 1$

> In a tree the number of edges is always one less than the number of nodes.

a For example
- In Prim the tree 'grows' is a connected fashion.
- There is no need to check for cycles when using Prim.
- Prim can be adapted to matrix form.
- Prim starts with a vertex, Kruskal with an edge.
- Kruskal must start with the least edge, Prim can start with any vertex.

b i For example AC, AB, BD, EB, EF, EG   **ii** EF, AC, BD, AB, reject AD, EG, reject FG, BE

13 a For example, 45 37 18 $\boxed{46}$ 56 79 90 81 51

b For example, 56 45 79 46 37 90 81 51 18

c 1st pivot $\left[\dfrac{1+11}{2}\right] \to$ 6th number (44) $73 > 44$ so reject 1st to 6th numbers

2nd pivot $\left[\dfrac{7+11}{2}\right] \to$ 9th number (71) $73 > 71$ so reject 7th to 9th numbers

3rd pivot $\left[\dfrac{10+11}{2}\right] \to$ 11th number (94) $73 < 94$ so reject 11th number

4th pivot 10th number (73) item found   73 was found as the 10th number in the list.

14 a For example,

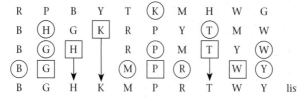

list is order

b $\left[\dfrac{10+1}{2}\right] \to$ 6. Palmer   Kenney < Palmer reject 6 to 10

list is now
1 Boase
2 Garesalingham
3 Halliwell
4 Kenney
5 Morris

2nd pivot $\left[\dfrac{1+5}{2}\right] \to$ 3. Halliwell   Kenney > Halliwell reject 1 to 3

list is now
4 Kenney
5 Morris

3rd pivot $\left[\dfrac{4+5}{2}\right] \to$ 5. Morris   Kenney < Halliwell reject 5

list is now
4 Kenney
item found
Kenney was found as the 4th name in the list.

15 a The list is not in alphabetical order.

b For example, quick sort.

c
1 Birmingham
2 Cardiff
3 Exeter
4 Glasgow
5 Leicester
6 Manchester
7 Newcastle
8 Plymouth
9 Southampton
10 York

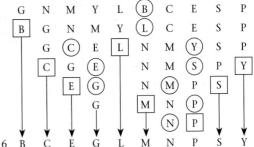

$\left[\dfrac{1+10}{2}\right] \to 6\,$Manchester   Newcastle > Manchester so reject 1 to 6

list is now:
7 Newcastle
8 Plymouth
9 Southampton
10 York

$\left[\dfrac{7+10}{2}\right] \to 9\,$Southampton   Newcastle < Southampton so reject 9 to 10

list is now
7 Newcastle
8 Plymouth

$\left[\dfrac{7+8}{2}\right] \to 8\,$Plymouth   Newcastle < Plymouth so reject 8

list is now
7 Newcastle
Final term is 7, Newcastle ∴ name found at 7.

16 a

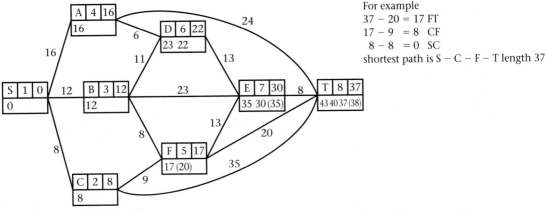

For example
37 − 20 = 17 FT
17 − 9 = 8 CF
8 − 8 = 0 SC
shortest path is S − C − F − T length 37

b Need shortest path S to E + ET
So S − C − F − E − T length 38

17 a

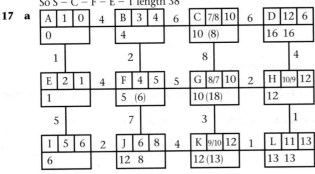

For example
13 − 1 = 12 HL   or   13 − 1 = 12 KL
12 − 2 = 10 GH   12 − 4 = 8 JK
10 − 5 = 5 FG   8 − 2 = 6 IJ
5 − 4 = 1 EF   6 − 5 = 1 EI
1 − 1 = 0 AE   1 − 1 = 0 AE

Shortest path is $\left\{\begin{array}{l}A - E - F - G - M - L\\ A - E - I - J - K - L\end{array}\right\}$ length 13

b State the other path.

18 a BD + FG = 1.3 + 0.9 = 2.2
BF + DG = 1.5 + 2.0 = 3.5   ———————— The odd valencies are at F, B, D, F, G.
BG + DF = 0.7 + 1.7 = 2.4
Repeat BD and FG
Route for example G − F − E − D − B − C − D − B − F − G − A − B − G   (13 letters in route)
length 8.9 + 2.2 = 11.1 km

b It would now only be necessary to repeat BF of length 1.5 < 2.2 length = 8.9 + 1.5 = 10.4 km, saving 0.7 km.

19 a

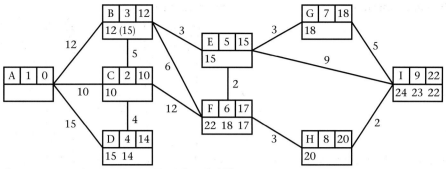

Shortest route is A − B − E− F − H − I, length 22 km

b i Odd vertices are A and I (only), so we need to repeat the shortest route from A to I. This was found is **a**.
so repeat A B, B E, E F, F H, H I.
ii For example A − B − C − A − D − C − E − H − I − H − E − F − I − G − F − E − B − F − B − A (20 letters in route)
iii 91 + 22 = 113 km

20 a For example bubbling left to right

90	50	55	40	20	35	30	25	45
90	55	50	40	35	30	25	45	20
90	55	50	40	35	30	45	25	20
90	55	50	40	35	45	30	25	20
90	55	50	40	45	35	30	25	20
90	55	50	45	40	35	30	25	20

No more changes, list sorted.

b $\frac{475}{120}$ = 3.96 so lower bound is 4 tapes

c Bin 1: 90 + 30
Bin 2: 55 + 50
Bin 3: 45 + 40 + 35
Bin 4: 35 + 30 + 25 + 20
Bin 5: 20

d For example
Bin 1: 90 + 30
Bin 2: 55 + 35 + 30
Bin 3: 45 + 40 + 35
Bin 4: 50 + 25 + 20 + 20

21 a i Shortest path through A is 18 + y or 26 both of which are greater than 17.
Shortest path through C is 23, which is greater than 17. So shortest path cannot go through A or C.
ii Shortest path must go through *B*.
$$S − B − D − T = 13 + x$$
$$13 + x = 17$$
$$x = 4$$

b If $y = 0$ shortest path is S − A − D − T = 18   If $y = 5$ shortest path is S − C − D − T = 23   so range is 18 to 23.

c For example, a person seeking the quickest route from home to work through a city. The arcs are the roads that may be chosen, the number the time, in minutes, to journey along that road. The nodes represent junctions.

22 a i A connected graph with no cycles, loops or multiple edges.
ii A tree that includes all vertices.
iii A spanning tree of minimum total weight.

b For example,
- There is no need to check for cycles when using Prim's algorithm.
- Prim's algorithm can start at any vertex, Kruskal's algorithm starts with the shortest arc.
- In Prim's algorithm the tree 'grows' in a connected fashion, with Kruskal's algorithm the tree may not be connected until the end.
- When using Kruskal's algorithm the shortest *arc* is added to the tree (unless it completes a cycle); with Prim's algorithm the nearest unattached *vertex* is added.

c BH, NF, HN, AH, BE, AN, EF, AF, HE, CN, CE, BC, EN
Use BH, NF, HN, AH, BE, reject NA, reject EF, reject AF, reject HE, CN

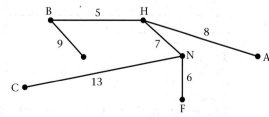

d

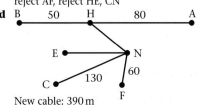

New cable: 390 m

23 a 75   70   60   50   40   35   20   20   20

Bin 1: 75 + 20
Bin 2: 70 + 20
Bin 3: 60 + 40
Bin 4: 50 + 35
Bin 5: 20
5 planks needed at a cost of 5 × £3 = £15
wastage is 5 + 10 + 0 + 15 + 80 = 110 cm

   b For example

Bin 1 (1.5 m): 75 + 70   *or*   Bin 1 (1 m): 75 + 20
Bin 2 (1.5 m): 60 + 50 + 40   Bin 2 (1.5 m): 70 + 60 + 20
Bin 3 (1 m): 35 + 20 + 20 + 20   Bin 3 (1.5 m): 50 + 40 + 35 + 20
Cost: 2 × £4 + £3 = £11
1.5 m lengths are better value than 1 m lengths, therefore the solution using as many 1.5 m lengths as possible is preferred.

24 a

Route: S – A – C – G – T length: 82

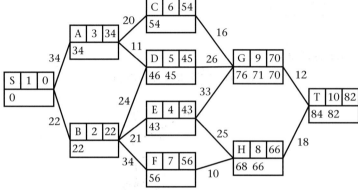

   b For example   82 − 12 = 70 GT
                      70 − 16 = 54 CG
                      54 − 20 = 34 AC
                      34 − 34 = 0 SA

   c Shortest route from S to H + HT   S − B − F − H − T length: 84

25 a

a	*b*	*c*	*d*	*e*	*f*	*f* = 0?
645	255	2.53	2	510	135	No
255	135	1.89	1	135	120	No
135	120	1.13	1	120	15	No
120	15	8	8	120	0	Yes

answer is 15

   b The first row would be
    255   645   0.40   0   0   255   No
    but the second row would then be the same as the first row in the table above. So in effect this new first line would just be an additional row at the start of the solution.

   c Finds the Highest Common Factor of *a* and *b*.

26 a For example, left to right

55   80   25   84   25   34   17   75   3   5
80   55   84   25   34   25   75   17   5   3
80   84   55   34   25   75   25   17   5   3
84   80   55   34   75   25   25   17   5   3
84   80   55   75   34   25   25   17   5   3
84   80   75   55   34   25   25   17   5   3
No more exchanges, sort complete.

   b 403 ÷ 100 = 4.03 so 5 bins are needed.

   c Bin 1: 84 + 5 + 3
Bin 2: 80 + 17
Bin 3: 75 + 25
Bin 4: 55 + 34
Bin 5: 25

27 a

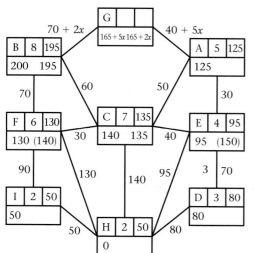

Via A   H − E − A − G   length $165 + 5x$

Via B   H − E − C − B − G   length $265 + 2x$

b $165 + 5x < 265 + 2x \Rightarrow x , 33\frac{1}{3}$ So range is $0 \leqslant x < 33\frac{1}{3}$

28 a

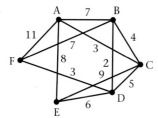

b   BD, $\begin{Bmatrix} AC \\ DF \end{Bmatrix}$, BC, reject CD, DE

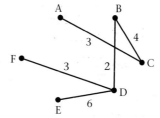

c   DB, DF, BC, CA, DE length of tree 18 km

29 a   Odd vertices are B_1, B_2, E, G

    $B_1 B_2 + EG = 65 + 18 = 83$

    $B_1 E + B_2 G = 41 + 42 = 83$

    $B_1 G + B_2 E = 26 + 30 = 56$

    Repeat BD, DG, B_2A, AE

    Route: For example,   F − A − E − A − B_2 − A − C − E − F − G − D − H − G − D − B_1 − D − F

    (All correct routes have 17 letters in their 'word')

    length $= 129 + 56 = 185$ km

b   Now only the route between E and G needs repeating so repeat EF + FG = 18

    length of new route $= 129 + 18$

                      $= 147$ km

30 a

a	b	c	Integer?	Output list	$a = b$?
90	2	45	Y	2	N
45	2	22.5	N		
45	3	15	Y	3	N
15	2	7.5	N		
15	3	5	Y	3	N
5	2	2.5	N		
5	3	$1\frac{2}{3}$	N		
5	5	1	Y	5	Y

b   Expresses a as a product of prime factors

c $c = 1 \left(\text{since we stop when } a = b \text{ and } c = \frac{a}{b} \right)$

Output list: 2, 3, 3, 5

31 a i

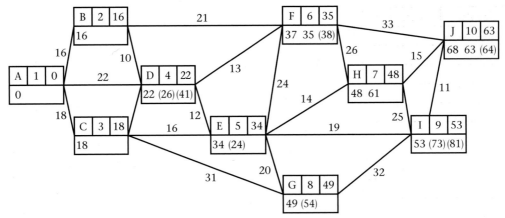

ii For example

63 − 15 = 48	HJ	
48 − 14 = 34	EH	*or*
34 − 16 = 18	CE	
18 − 18 = 0	AC	

63 − 15 = 48	HJ
48 − 14 = 34	EH
34 − 12 = 22	DE
22 − 22 = 0	AD

Route A − C − E − H − J or A − D − E − H − J length 63

b No, it is not the only route, there are two shortest routes (see above).

Exercise 5A

1 One possible solution is:

Activity	Depends on
A	—
B	A
C	· B
D	C
E	D
F	E
G	F

Another possible solution is:

Activity	Depends on
A	—
B	A
C	—
D	B, C
E	D
F	E
G	F

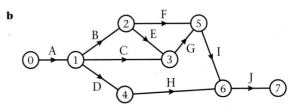

You can start the engine with the clutch depressed.

2 a

Activity	Depends on
A	—
B	A
C	A
D	A
E	B
F	B
G	C, E
H	D
I	F, G
J	H, I

b

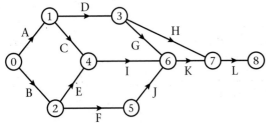

3

4

Activity	Depends on
A	–
B	A
C	A
D	B
E	C
F	E
G	D, F
H	E
I	E
J	H
K	I

Exercise 5B

1

Activity	Depends on
A	—
B	—
C	A
D	A
E	C
F	B, D
G	B

The dummy shows that activity F depends on activities B and D, whereas activity G only depends on activity B.

2

Activity	Depends on
A	—
B	—
C	A
D	A
E	C
F	B, C, D
G	B, C, D
H	E, F

3

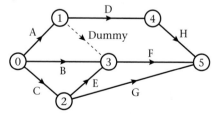

4

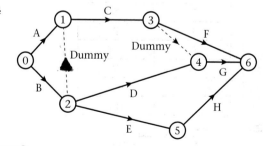

5

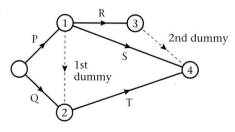

1st dummy ②.
S depends on P only.
T depends on P and Q.
2nd dummy.
So that S and R don't share a start and end event.

Exercise 5C

1 x is the largest of $7 + 5 = 12$, $5 + 8 = 13$ and $9 + 5 = 14$. $x = 14$

2

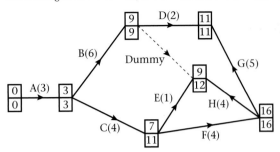

$w = 11, x = 16,$
$y = 11, z = 9$

> w and x are found using a forward scan.
> y and z are found using a backward scan.

3

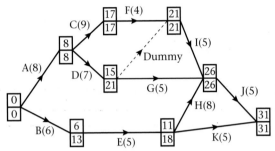

Exercise 5D

1 $x = 3$, $y = 17$, $z = 17$
2 a The critical activities are B, E, H, J and N   **b** I
3 a   **b** No. $7 + 1 \neq 14$

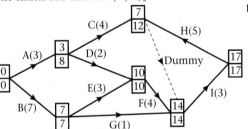

c The critical path is B – E – F – I.

Exercise 5E

1

Activity	Total float
A	0
B	$10 - 3 - 0 = 7$
C	$15 - 8 - 6 = 1$
D	0
E	$14 - 4 - 3 = 7$
F	$20 - 5 - 14 = 1$
G	0
H	$22 - 8 - 7 = 7$
I	$28 - 8 - 19 = 1$
J	$22 - 2 - 19 = 1$
K	$29 - 1 - 27 = 1$
L	0

2 a $a = 10$ $b = 19$ $x = 19 - 10 = 9$
  Total float $= 3 = 15 - y - a$
        $y = 15 - 3 - 10$
        $y = 2$

b Minimum value of $c = 10 + 2 = 12$
c Minimum value of total float of R $= 19 - 5 - 12$
        $= 2$

Exercise 5F

1

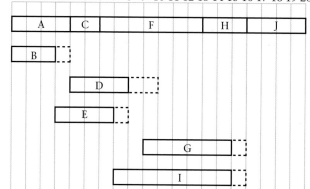

2 a $w = 10$ $x = 19$ $y = 19$ $z = 19$ **b** Critical activities: B, E, H, M, O

  c Total float for E $= 26 - 5 - 13 = 8$ Total float for N $= 39 - 5 - 29 = 5$

  d

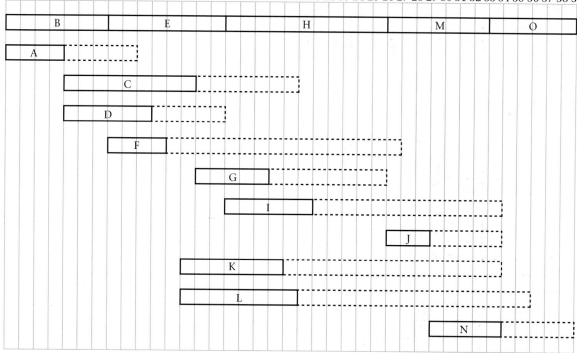

Exercise 5G

1 a C, D, E, F   **b** J, I   **c** J, F, H, J

2 a C, D   **b** E, G

3   0 1 2 3 4 5 6 7 8 9 10 11 12 13 14   **a** A, C, D

                                                      b E, C

Exercise 5H

1   **a** $\frac{64}{22} = 2.9\ldots$ so, lower bound = 3

    b No. 2 hours is less than the total float for activity B (3 hours).

    c Activity H

    d 0 1 2 3 4 5 6 7 8 9 10 11 12 13 14 15 16 17 18 19 20 21 22

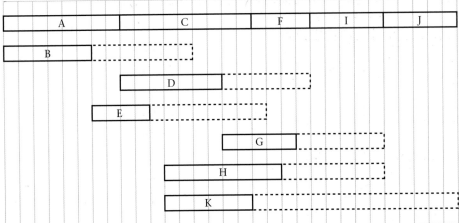

    4 workers are needed to complete the project in 22 hours.

2   **a** 0 1 2 3 4 5 6 7 8 9 10 11 12 13 14 15 16 17 18 19 20 21 22 23 24 25 26 27 28 29 30 31

(Gantt chart: A, C, F, I, J; B; D; E; G; H; K)

    b 0 1 2 3 4 5 6 7 8 9 10 11 12 13 14 15 16 17 18 19 20 21 22 23 24 25 26 27 28 29 30 31

(Scheduling chart: A, C, F, I, J; B, E, D, G, K; D)

    3 workers are needed to complete the project in the critical time.

3 0 1 2 3 4 5 6 7 8

(Scheduling chart: A; B, E)

> When worker 2 completes activity B, only activity E may be started.

0 1 2 3 4 5 6 7 8 9 10 11 12 13

(Scheduling chart: A, C, D; B, E)

> When worker 1 completes activity A, the next activity to start is either C or D. Activity C is chosen because it has the lower value for its latest finish time.

0 1 2 3 4 5 6 7 8 9 10 11 12 13 14 15

(Scheduling chart: A, C, D; B, E, F)

> At this stage, worker 1 has a choice between activity G and activity I. Activity H may not be started until activity F has been completed.

There are two possible ways to complete the schedule so that the project is completed in the minimum possible time.

```
0 1 2 3 4 5 6 7 8 9 10 11 12 13 14 15 16 17 18 19 20 21 22 23 24 25 26
```

A	C	D		G		H	
B	E		F		I		J

```
0 1 2 3 4 5 6 7 8 9 10 11 12 13 14 15 16 17 18 19 20 21 22 23 24 25 26
```

A	C	D		I		H	J
B	E		F		G		

Mixed Exercise 5I

1 a Activity D depends on activities A and C, whereas activity E depends only on activity A.
This shows that a dummy is required.
Activity J depends on activities G and I, whereas activity H depends only on activity G.
This shows that a second dummy is required.

b

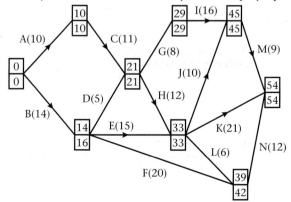

2 a

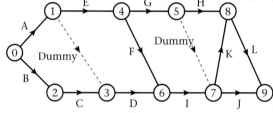

b Dummy 1 is needed to show *dependency*.
E and F depend on C and B, but D depends on B only.
Dummy 2 is need so that each activity can be uniquely represented in terms of its event.

3 a

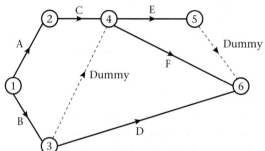

b There are *two* critical paths:
A – C – G – I – M and A – C – H – K – M
The critical activities are A, C, G, H, I, K

c Total float on D is 21 − 5 − 14 = 2
Total float on F is 42 − 20 − 14 = 8

d 0 2 4 6 8 10 12 14 16 18 20 22 24 26 28 30 32 34 36 38 40 42 44 46 48 50 52 54

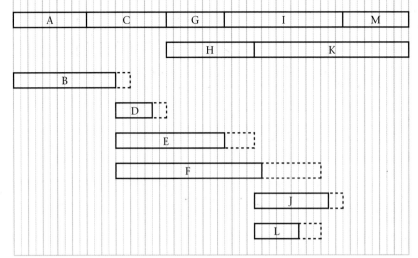

e Day 15: C
Day 25: G, H, E, F

4 a J depends on H alone, but L depends on H and I

b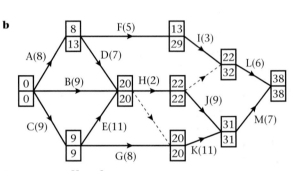

c Total float on D = 20 − 7 − 8 = 5
Total float on E = 20 − 11 − 9 = 0
Total float on F = 29 − 5 − 8 = 16

d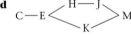

C — E < H — J
          K > M

e $\frac{95}{38}$ = 2.5 so 3 workers

f For example   0 2 4 6 8 10 12 14 16 18 20 22 24 26 28 30 32 34 36 38 40

C		E		H		J		M	
B		G			K				
A		D		F		I		L	

A < F / D   C < E / G   B / D / E > H   B / D / E / G > K   J / K > M   H / I > L

5 a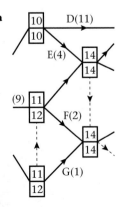

b The critical paths are: A − E − H − K and A − E − L

c A critical path is a continuous path from the source node to the sink node such that a delay in any activity results in a corresponding delay in the whole project.

d $\frac{\text{Sum of all of the activity times}}{\text{critical time of the project}} = \frac{110}{30}$

Lower bound for number of workers is 4.

e D, H, I, J, L

f The answers to part **e** show that 5 workers are needed on day 20 in order to complete the project in the minimum time.

g 0 1 2 3 4 5 6 7 8 9 10 11 12 13 14 15 16 17 18 19 20 21 22 23 24 25 26 27 28 29 30

A		E		H		K	
B		F		I	
C		G		J	
	D		
	L	

Exercise 6A

1 Number of boxes of gold assortment $= x$ Number of boxes of silver assortment $= y$
Objective: maximise $P = 80x + 60y$
Constraints
- time to make chocolate, $30x + 20y \leq 300 \times 60$ which simplifies to $3x + 2y \leq 1800$
- time to wrap and pack $\quad 12x + 15x \leq 200 \times 60$ which simplifies to $4x + 4x \leq 4000$ •——— All units now in minutes.
- 'At *least* twice as many silver as gold' $2x \leq y$
- non-negativity $x, y \leq 0$
In summary: maximise $P = 80x + 60y$
subject to
$3x + 2y \leq 1800$
$4x + 5y \leq 4000$
$\quad 2x \leq y$
$\quad x, y \geq 0$

2 number of type A $= x$ number of type B $= y$
Objective: minimise $c = 6x + 10y$
Constraints
- Display must be at least 30 m long $x + 1.5y \geq 30$ which simplifies to $2x + 3y \geq 60$
- 'At least twice as many x as y' $2y \leq x$
- At *least* six type B $y \geq 6$
- non-negativity $x, y \geq 0$
In summary: minimise $c = 6x + 10y$
subject to:
$2x + 3y \geq 60$
$\quad 2y \leq x$
$\quad\quad y \geq 6$
$\quad x, y \geq 0$

3 Number of games of Cludopoly $= x$ Number of games of Trivscrab $= y$
Objective: maximise $P = 1.5x + 2.5y$
Constraints
- First machine: $\quad 5x + 8y \leq 10 \times 60$ which simplifies to $5x + 8y \leq 600$
- Second machine: $\quad 8x + 4y \leq 10 \times 60$ which simplifies to $2x + y \leq 150$ •——— All units now in minutes.
- At *most* 3 times as many x as y $3y \geq x$
- non-negativity $x, y \geq 0$.
In summary: maximise $P = 1.5x + 2.5y$
subject to:
$5x + 8y \leq 600$
$2x + y \leq 75$
$\quad 3y \geq x$
$\quad x, y \geq 0$

4 Number of type 1 bookcases $= x$ Number of type 2 bookcases $= y$
Objective: maximise $S = 40x + 60y$
Constraints
- budget $\quad 150x + 250y \leq 3000$ which simplifies to $3x + 5y \leq 60$
- floor space $\quad 15x + 12y \leq 240$ which simplifies to $5x + 4y \leq 80$
- 'At most $\frac{1}{3}$ of all bookcases to be type 2' $y \leq \frac{1}{3}(x + y)$ which simplifies to $2y \leq x$
- At least 8 type 1 $\quad x \geq 8$
- non-negativity $x, y \geq 0$
In summary: maximise $S = 40x + 60y$
subject to:
$3x + 5y \leq 60$
$5x + 4y \leq 80$
$\quad 2y \leq x$
$\quad\quad x \geq 8$
$\quad x, y \geq 0$

5 Let x = number of kg of indoor feed and y = number of kg of outdoor feed

Objective: maximise $P = 7x + 6y$

Constraints

- Amount of A $10x + 20y \leqslant 5 \times 1000$ which simplifies to $x + 2y \leqslant 500$
- Amount of B $20x + 10y \leqslant 5 \times 1000$ which simplifies to $2x + y \leqslant 500$ •——— All units now in grams.
- Amount of C $20x + 20y \leqslant 6 \times 100$ which simplifies to $x + y \leqslant 300$
- At *most* 3 times as much y as x $y \leqslant 3x$
- At least 50 kg of x $x \geqslant 50$
- non-negativity $y \geqslant 0$ ($x \geqslant 0$ is unnecessary because of the previous constraint)

In summary: maximise $P = 7x + 6y$

subject to

$x + 2y \leqslant 500$

$2x + y \leqslant 500$

$x + y \leqslant 300$

$y \leqslant 3x$

$x \geqslant 50$

$y \geqslant 0$

6 number of A smoothies $= x$ number of B smoothies $= y$ number of C smoothies $= z$

Objective: maximise $P = 60x + 65y + 55z$

Constraints

- oranges $x + \frac{1}{2}y + 2z \leqslant 50$ which simplifies to $2x + y + 4z \leqslant 100$
- raspberries $10x + 40y + 15z \leqslant 1000$ which simplifies to $2x + 8y + 3z \leqslant 200$
- kiwi fruit $2x + 3y + 2 \leqslant 100$
- apples $2x + \frac{1}{2}y + 2z \leqslant 60$ which simplifies to $4x + y + 4z \leqslant 120$
- non-negativity $x, y, z \geqslant 0$

In summary: maximise $P = 60x + 65y + 55z$

subject to:

$2x + y + 4z \leqslant 100$

$2x + 8y + 3z \leqslant 200$

$2x + 3y + z \leqslant 100$

$4x + y + 4z \leqslant 120$

$x, y, z \geqslant 0$

7 Let number of hours of work for factory R $= x$ Let number of hours of work for factory S $= y$

Objective: minimise $C = 300x + 400y$

Constraints

- milk $1000x + 800y \geqslant 20\,000$ which simplifies to $5x + 4y \geqslant 100$
- yoghurt $200x + 300y \geqslant 6000$ which simplifies to $2x + 3y \geqslant 60$
- At *least* $\frac{1}{3}$ of total time for R $x \geqslant \frac{1}{3}(x + y)$ which simplifies to $2x \geqslant y$
- At *least* $\frac{1}{3}$ of total time for S $y \geqslant \frac{1}{3}(x + y)$ which simplifies to $2y \geqslant x$
- non-negativity $x, y \geqslant 0$

In summary: minimise $C = 300x + 400y$

subject to:

$5x + 4y \geqslant 100$

$2x + 3y \geqslant 60$

$2x \geqslant y$

$2y \geqslant x$

$x, y \geqslant 0$

Exercise 6B

1

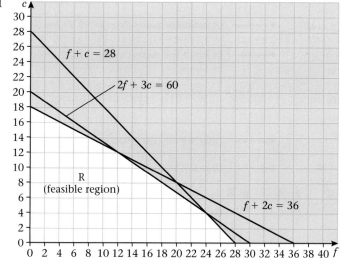

4

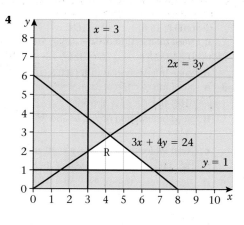

2

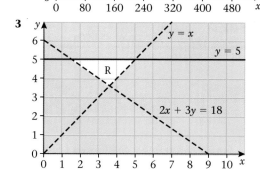

5

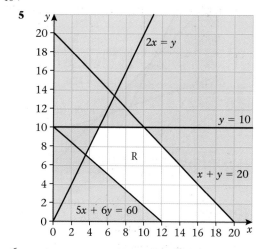

3

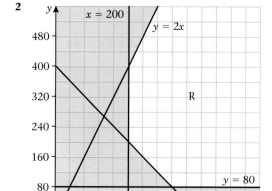

6

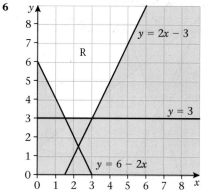

Exercise 6C

1 **a** Need intersection of $4x + y = 1400$ and $3x + 2y = 1200$ ———— Objective line passes through (200, 0) and (0, 400).
  (320, 120) $m = 760$

b (0, 400) $N = 1600$ ———— Objective line passes through (400, 0) and (100, 0).

c Need intersection of $x + 3y = 1200$ and $3x + 2y = 1200$ ———— Objective line passes through (200, 0) and (0, 200).
  $\left(171\frac{3}{7}, 342\frac{6}{7}\right)$ $P = 514\frac{2}{7}$

d (350, 0) $Q = 2100$ ———— Objective line passes through (100, 0) and (0, 600).

2 a $(0, 90)$ $E = 90$

  b Need intersection of $6y = x$ and $3x + 7y = 420$
     $(100.8, 16.8)$ $F = 168$

  c Need intersection of $9x + 10y = 900$ $3x + 7y = 420$
     $\left(63\frac{7}{11}, 32\frac{8}{11}\right)$ $G = 321\frac{9}{11}$

  d Same intersection as in **b** $(100.8, 16.8)$ $H = 201.6$

3 a Need intersection of $3x + y = 60$ and $5y = 3x$
     $\left(16\frac{2}{3}, 10\right)$ $J = 56\frac{2}{3}$

  b Need intersection of $y = 4x$ and $9x + 5y = 45$
     $\left(1\frac{16}{29}, 6\frac{6}{29}\right)$ $K = 7\frac{22}{29}$

  c Need intersection of $3x + y = 60$ and $y = 4x$
     $\left(8\frac{4}{7}, 34\frac{2}{7}\right)$ $L = 65\frac{5}{7}$

  d Need intersection of $9x + 5y = 450$ and $5y = 3x$
     $(37.5, 22.5)$ $m = 97.5$

• Objective line passes through $(80, 0)$ and $(0, 60)$.

• Objective line passes through $(120, 0)$ and $(0, 20)$.

• Objective line passes through $(10, 0)$ and $(0, 60)$.

• Objective line passes through $(40, 0)$ and $(0, 80)$.

4 a C   **b** A   **c** B   **d** D   **e** C   **f** A   **g** B   **h** D   **i** C   **j** D

5 a

b

Vertices	$C = 3x + 2y$
$(0, 160)$	320
$(40, 80)$	280
$(90, 30)$	330
$(180, 0)$	540

so minimum is $(40, 80)$ value of $C = 280$

c $(90, 30)$ $C_1 = 270$

d C_2 is parallel to $x + y = 120$ so all points from A to B are optimal points.

6 a

b i $(12, 6)$ $P = 30$

   ii $\left(34\frac{2}{37}, 17\frac{1}{37}\right)$ $Q = 221\frac{13}{37}$

7 a

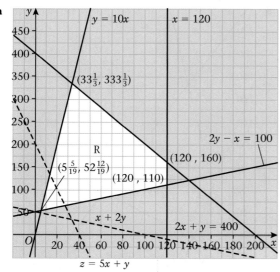

$y = 10x$ · $x = 120$

$(33\frac{1}{3}, 333\frac{1}{3})$

$2y - x = 100$

R

$(120, 160)$

$(5\frac{5}{19}, 52\frac{12}{19})$

$(120, 110)$

$x + 2y$

$2x + y = 400$

$z = 5x + y$

b i $(120, 100)$ $z = 700$

  ii $\left(5\frac{5}{19}, 52\frac{12}{19}\right)$ $z = 78\frac{18}{19}$

c Optimal point $\left(33\frac{1}{3}, 333\frac{1}{3}\right)$ optimal value of $x + 2y = 700$

8

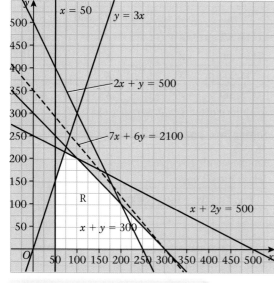

$x = 50$ $y = 3x$

$2x + y = 500$

$7x + 6y = 2100$

$x + 2y = 500$

R

$x + y = 300$

Objective line passes through (0, 350) and (300, 0). Maximum point is (300, 0).

9

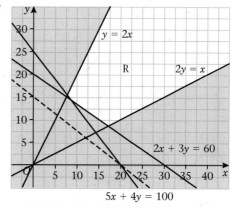

$y = 2x$

R $2y = x$

$2x + 3y = 60$

$5x + 4y = 100$

Objective line passes through (0, 15) and (20, 0). Intersection of $2x + 3y = 60$ $5x + 4y = 100$ $\left(8\frac{4}{7}, 14\frac{2}{7}\right)$ value $= 8285\frac{5}{7}$

Exercise 6D

1

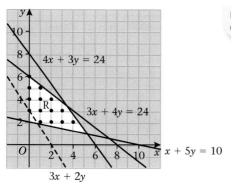

$4x + 3y = 24$

R

$3x + 4y = 24$

$x + 5y = 10$

$3x + 2y$

Maximum integer value (5, 1) $3x + 2y = 17$

2

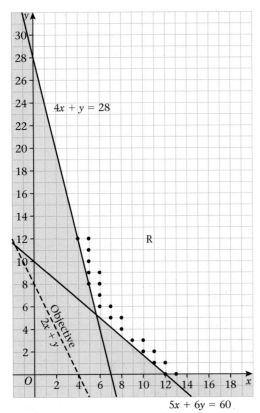

$4x + y = 28$

R

Objective
$2x + y$

$5x + 6y = 60$

Minimum integer value (6, 6) $2x + y = 18$

3

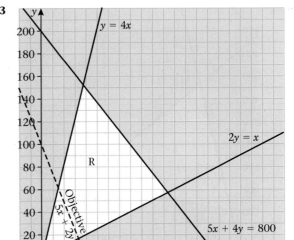

$y = 4x$

$2y = x$

R

Objective
$5x + 2y$

$5x + 4y = 800$

Solving $2y = x$ and $5x + 4y = 800$ simultaneously gives $\left(114\frac{2}{7}, 57\frac{1}{7}\right)$

Test integer values nearby.

Point	$2y \geqslant x$	$5x + 4y \leqslant 800$	$5x + 2y$
(114, 57)	✓	✓	684
(114, 58)	✓	✗	—
(115, 57)	✗	✗	—
(115, 58)	✓	✗	—

so optimal point is (114, 57) value 684

Solving $3x + 16y = 2400$
$3x + 5y = 1500$
simultaneously gives $\left(363\frac{7}{11}, 81\frac{9}{11}\right)$
Taking integer point

Point	$3x + 16y \geqslant 2400$	$3x + 5y \leqslant 1500$
(363, 81)	✓	✗
(363, 82)	✓	✓
(364, 81)	✓	✗
(364, 82)	✗	✓

So optimal integer point is (363, 82)
Value 808

4

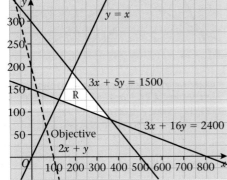

$y = x$

$3x + 5y = 1500$

R

$3x + 16y = 2400$

Objective
$2x + y$

Intersection of $4x + 5y = 4000$ and $3x + 2y = 1800$
gives $\left(142\frac{6}{7}, 685\frac{5}{7}\right)$
Testing nearby integer points

Point	$4x + 5y \leqslant 4000$	$3x + 2y \leqslant 1800$	$4x + 3y$
(142, 685)	✓	✓	2653
(142, 686)	✓	✓	2646
(143, 685)	✓	✓	2627
(143, 686)	✗	✗	—

so maximum integer solution is 2627 at (143, 685).

5

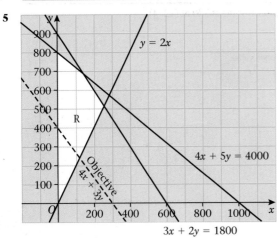

$y = 2x$

R

$4x + 5y = 4000$

Objective
$4x + 3y$

$3x + 2y = 1800$

6

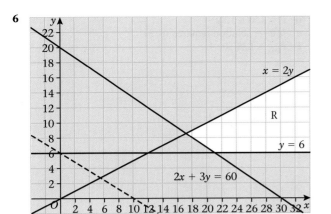

Objective
$6x + 10y$

Intersection of $2x + 3y = 60$ and $y = 6$
$(21, 6)$
$$\text{cost } c = 6x + 10y$$
$$\text{so minimum cost} = 6 \times 21 + 60$$
$$= \text{£}186$$

7

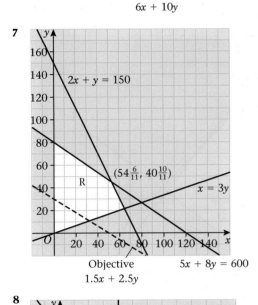

Objective $5x + 8y = 600$
$1.5x + 2.5y$

Intersection of $5x + 8y = 600$
$2x + y = 150$ giving $\left(54\frac{6}{11}, 40\frac{10}{11}\right)$

Points	$5x + 8y \leq 600$	$2x + y \leq 150$	$1.5x + 2.5x$
$(54, 40)$	✓	✓	181
$(54, 41)$	✓	✓	183.5
$(55, 40)$	✓	✓	182.5
$(55, 41)$	✗	✗	—

so maximum value is 183.5 at $(54, 41)$.

8

Objective line $5x + 4y = 80$
$40x + 60y$

Intersection of $3x + 5y = 60$
$5x + 4y = 80$ giving $\left(12\frac{4}{13}, 4\frac{8}{13}\right)$

Points	$3x + 5y \leq 60$	$5x + 4y \leq 80$	$40x + 60y$
$(12, 4)$	✓	✓	720
$(12, 5)$	✗	✓	—
$(13, 4)$	✓	✗	—
$(13, 5)$	✗	✗	—

Maximum value is 720 at $(12, 4)$.

Mixed exercise 6E

1 a *flour*: $200x + 200y \leq 2800$ so $x + y \leq 14$
    fruit: $125x + 50y \leq 1000$ so $5x + 2y \leq 40$
  b Cooking time $50x + 30y \leq 480$ so $5x + 3y \leq 48$

c

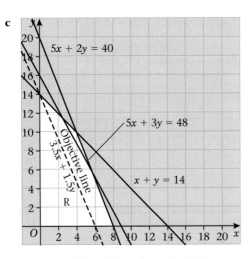

d $P = 3.5x + 1.5y$
e Integer solution required (6, 5)
f $P_{max} = £28.50$

---

2 b cost: $6x + 4.8y \leqslant 420$ $5x + 4y \leqslant 350$
c Display $30x \leqslant 2 \times 20y$ $3x \leqslant 4y$
e Integer solution required (43, 33).

d

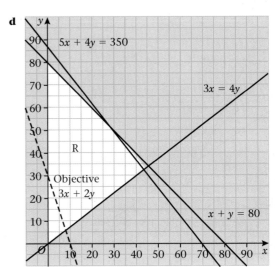

---

3 a i Total number of people
$54x + 24y \geqslant 432$ so $9x + 4y \geqslant 72$
ii number of adults $x + y \leqslant 12$
iii number of large coaches $x \leqslant 7$
c minimise $C = 336x + 252y$
$= 84(4x + 3y)$
d Objective line passes through (0, 4) (3, 0)
e Integer coordinates needed (7, 3)
cost = £3108

b

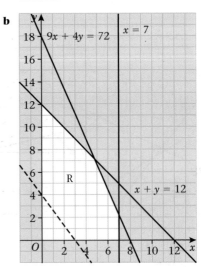

Objective
$84(4x + 3y)$

4 a $4x + 5y \leqslant 47$
$y \geqslant 2x - 8$
$4y - y - 8 \leqslant 0$
$x, y \geqslant 0$

b Solving simultaneous equations $y = 2x - 8$
$4x + 5y = 47$
$6\frac{3}{14}, 4\frac{3}{7}$

c i For example where x and y
  • types of car to be hired
  • number of people etc

ii (6, 4)

5 a $2\frac{1}{2}x + 3y \leqslant 300 \ [5x + 6y \leqslant 600]$
$5x + 2y \leqslant 400$
$2y \leqslant 150 \ [y \leqslant 75]$

b Maximise $P = 2x + 4y$

c (30, 75) $P = 360$

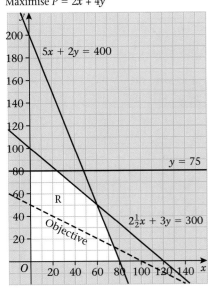

d The optimal point is at the intersection of $y = 75$ and $2\frac{1}{2}x + 3y = 300$.
So the constraint $5x + 2y \leqslant 400$ is not at its limit.
At (30, 75) $5x + 2y = 300$ so 100 minutes are unused.

Exercise 7A

1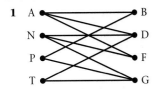

2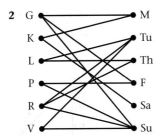

3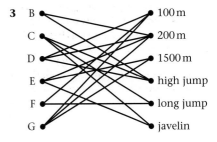

Mixed Exercise 7B

1 a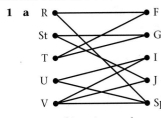

bipartite graph

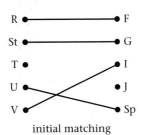

initial matching

either
Alternating path: T − G = St − J
Change states T = G − St = J
Complete matching
R = F St = J J = G U = SP V = I

or
Alternating path: T − F = R − Sp = U − I = V − J
Change status T = F − R = Sp − U = I − V = J
complete matching
R = Sp St = G T = F U = I V = J

2 a

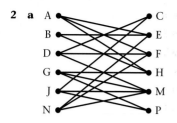 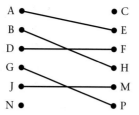

bipartite graph

initial matching

b,c Six possible alternating paths could be used, (one only is needed!)

i N – F = F – C

ii N – M = J – C

iii N – E = A – F = D – C

iv N – F = D – M = J – C

v N – E = A – F = D – M = J – C

vi N – M = J – P = G – H = B – E = A – F = D – C

Change status to give the complete matching

Person	Path (i)	Path (ii)	Path (iii)	Path (iv)	Path (v)	Path (vi)
A	E	E	F	E	F	F
B	H	H	H	H	H	E
D	C	F	C	M	M	C
G	P	P	P	P	P	H
J	M	C	M	C	C	P
N	F	M	E	F	E	M

3 a

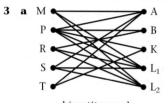 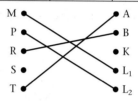

bipartite graph

initial matching

b Four possible alternating paths – one only is needed

i S – L = P – K

ii S – L = M – A = T – K

iii S – A = T – K

iv S – L = P – A = T – K

change status to give a complete matching

Person	Path (i)	Path (ii)	Path (iii)	Path (iv)
M	L	A	L	L
P	K	L	L	A
R	B	B	B	B
S	L	L	A	L
T	A	K	K	K

4 a 12 possible alternating paths – one only is needed

i F – 1 = B – 2 = A – 6

ii F – 1 = B – 2 = A – 4 = E – 6

iii F – 1 = B – 3 = D – 5 = C – 2 = A – 6

iv F – 1 = B – 3 = D – 5 = C – 2 = A – 4 = E – 6

v F – 1 = B – 4 = E – 6

vi F – 1 = B – 4 = E – 2 = A – 6

vii F – 5 = C – 2 = A – 1 = B – 4 = E – 6

viii F – 5 = C – 2 = A – 4 = E – 6

ix F – 5 = C – 2 = A – 6

x F – 3 = D – 5 = C – 2 = A – 6

xi F – 3 = D – 5 = C – 2 = A – 4 = E – 6

xii F – 3 = D – 5 = C – 2 = A – 1 = B = 4 – E = 6

Change status to give complete matching

Volunteer	(i)	(ii)	(iii)	(iv)	(v)	(vi)	(vii)	(viii)	(ix)	(x)	(xi)	(xii)
A	6	4	6	4	2	6	1	4	6	6	4	1
B	2	2	3	3	4	4	4	1	1	1	1	4
C	5	5	2	2	5	5	2	2	2	2	2	2
D	3	3	5	5	3	3	3	3	3	5	5	5
E	4	6	4	6	6	2	6	6	4	4	6	6
F	1	1	1	1	1	1	5	5	5	3	3	3

b Remove E and 2 and all arcs attached to each of them. Then run the algorithm as usual on the reduced problem.

5 a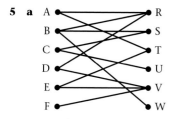

bipartite graph

b

initial matching

c There are 3 alternating paths that could be used

  i E – V = D – R or   **ii** E – T = A – R or   **iii** F – V = D – R

Change status to give

  i E = V – D = R or   **ii** E = T – A = R or   **iii** F = V – D = R

Maximum matching

Person	Path (i)	Path (ii)	Path (iii)
A	T	R	T
B	S	S	S
C	U	U	U
D	R	V	R
E	V	T	?
F	?	?	V

d For example C must do U and B must do W, but B and C are the only people who can do S. So a complete matching is not possible.

e

new bipartite graph

Depending on the path chosen in **c**
either E – V = D – R and F – V = E – T = A – S = B – W
or E – T = A – S = B – W and F – V = D – R
or E – T = A – S = B – R and F – V = D – R = B – W
Final matching
A = S B = W C = U E = T D = R F =

6 a A = Y B = X C = Z D = W E = V
A = X B = V C = W D = Y E = Z

b

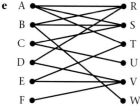

          V = E – Z = C – W
A ——— X = B – V = E – Z = C – W (given)
                   W (given)
  Y = D
               X = B – V = E – Z = C – W

  i Changing status in the two new paths
    A = V – E = Z – C = W and
    A = Y – D = X – B = V – E = Z – C = W
  ii Giving matchings
    A = V B = X C = W D = Y E = Z and
    A = Y B = V C = W D = X E = Z

7 a i

bipartite graph   initial matching

  ii

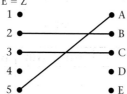

             A = 5 – D
      C = 3   A = 5 – D
1   E   4   E
    D   or   C = 3
                           A = 5 – D

So six possible alternating paths – one only needs to be used changing status

i 1 = D
ii 1 = C – 3 = E
iii 1 = C – 3 = A – 5 = D
iv 4 = A – 5 = D
v 4 = C – 3 = E
vi 4 = C – 3 = A – 5 = D

Giving improved matchings

Applicant	(i)	(ii)	(iii)	(iv)	(v)	(vi)
1	D	C	C	?	?	?
2	B	B	B	B	B	B
3	C	E	A	C	E	A
4	?	?	?	A	C	C
5	A	A	D	D	A	D

iii The next path depends on the previous path choice.

Previously chosen path	Possible next alternating path
i	4 – C = 3 – E
ii	4 – C = 1 – D
iii	4 – C = 1 – D = 5 – A = 3 – E
iv	1 – C = 3 – E
v	1 – D
vi	1 – D = 5 – A = 3 – E

Changing status leads to the following complete matching

Applicant	(i)	(ii)	(iii)	(iv)	(v)	(vi)
1	D	D	D	C	D	D
2	B	B	B	B	B	B
3	E	E	E	E	E	E
4	C	C	C	A	C	C
5	A	A	A	D	A	A

b If 3 matches with C

2 must now do E since no one else can but 2 is the only person who can do B too so a complete matching is not possible.

8 Mr + Mrs G must be assigned to two seats next to each other.

1 + 2 or 3 + 4 or 5 + 6 or 7 + 8

Depending on what you change we get

Alternative 1 (1 and 2 assigned to Mr + Mrs G)

Initially

A = 6
B = 5
C = 4
E = 7

For example, alternating paths

D – 7 = E – 8   and   F – 4 = C – 3

Change status D = 7 – E = 8 and F = 4 – C = 3

A = 6   C = 3   E = 8
B = 5   D = 7   F = 4

Mr + Mrs G in 1 and 2

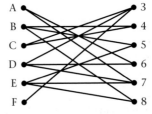

Alternative 2 (3 + 4 assigned to Mr + Mrs G)

Initially

A = 6
B = 5
D = 2
E = 7

For example, alternating paths

C – 2 = D – 7 = E – 8

Change status C = 2 – D = 7 – E = 8

A = 6   C = 2   E = 8
B = 5   D = 7   F = 1

Mr + Mrs G in 3 and 4

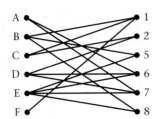

Alternative 3 (5 and 6 assigned to Mr + Mrs G)
Initially
C = 4
D = 2
E = 7
F = 1
For example, alternating paths
A − 7 = E − 8 and B − 4 = C − 3
Change status A = 7 − E = 8 and B = 4 − C = 3
A = 7   C = 3   E = 8
B = 4   D = 2   F = 1
Mr + Mrs in 5 and 6

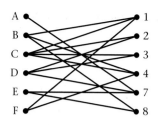

Alternative 4 (7 and 8 assigned to Mr + Mrs G)
Initially
A = 6
B = 5
C = 4
D = 2
F = 1
For example, alternating path
E − 5 = B − 4 = C − 3
Change status E = 5 − B = 4 − C = 3
A = 6   C = 3   E = 5
B = 4   D = 2   F = 1
Mr + Mrs G in 7 and 8

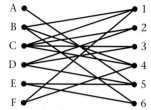

Review exercise 2

1　**a**　A graph consisting of
- two distinct sets of vertices X and Y in which
- arcs can only join a vertex in X to a vertex in Y.

　b　
- A path from an unmatched vertex in X to an unmatched vertex in Y
- which alternately uses arcs not in/in the matching.

> Two points to make in each of these responses.

　c　The one-to-one pairing of some elements of X with elements of Y.

　d　A one-to-one matching between all elements of X and Y.

2　**a**

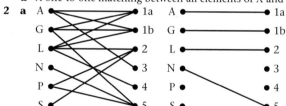

　b　Six possible alternating paths – only one needed
　　i　P − 2 = L − 4 change status P = 2 − L = 4
　　ii　P − 2 = L − 1 = G − 4 change status P = 2 − L = 1 − G = 4
　　iii　P − 2 = L − 1 = A − 3 change status P = 2 − L = 1 − A = 3
　　iv　S − 2 = L − 4 change status S = 2 − L = 4
　　v　S − 2 = L − 1 = G − 4 change status S = 2 − L = 1 − G = 4
　　vi　S − 2 = L − 1 = A − 3 change status S = 2 − L = 1 − A = 3
　　Giving matchings as follows

Person	(i)	(ii)	(iii)	(iv)	(v)	(vi)
A	1	1	3	1	1	3
G	1	4	1	1	4	1
L	4	1	1	4	1	1
N	5	5	5	5	5	5
P	2	2	2			
S				2	2	2

　c　For example, N must do 5 and S must do 2. This leaves P without a task.

3 a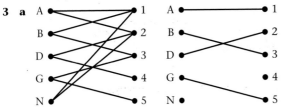

b Possible paths

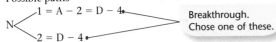

> Breakthrough.
> Chose one of these.

Change status

N = 1 – A = 2 – D = 4 giving matching A = 2   B = 3   D = 4   G = 5   N = 1

or

N = 2 – D = 4 giving matching A = 1   B = 3   D = 4   G = 5   N = 2

c Give the other alternating path.

4

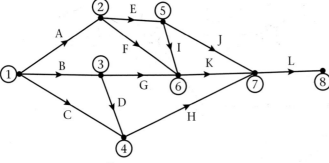

5 a

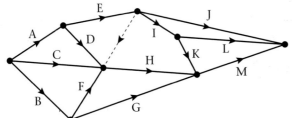

b For example I and J depend only on E.
H depends on C, D, E and F.

6 a

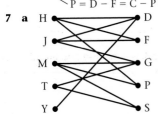

b Two possible paths – choose only one.

i B – T = E – V change status B = T – E = V

or

ii B – T = E – R change status B = T – E = R

Improved matching

i A = ? B = T C = F D = P E = V

or

ii A = ? B = T C = F D = P E = R

c For example, E is the only person who prefers V and the only person who prefers R. So a complete matching is not possible, so there cannot be a second alternating path.

or

A $\bigg\langle$ F = C – P = D – F loop
P = D – F = C – P loop

7 a

H •   • D
J •   • F
M •   • G
T •   • P
Y •   • S

b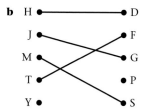

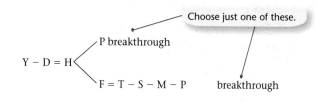

Change status either

i Y = D − H = P giving matching
   H = P   J = G   M = S   T = F   Y = D

or

ii Y = D − H = F − T = S − M = P giving matching
   H = F   J = G   M = P   T = S   Y = D

8 a

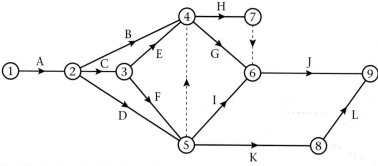

b D will only be critical if it lies on the longest path

Path A to G	Length
A − B − E −G	14
A − C − F − G	15
A − C − D == E − G	13 + x

So we need 13 + x to be the longest, or equal longest
13 + x ⩾ 15
      x ⩾ 2

9 a 2 − B = 4 − E change status 2 = B − 4 = E
   giving matching 1 = C   2 = B   3 = A   4 = E   5 = D
   2 − A = 3 − D = 5 − E change status 2 = A − 3 = D − 5 = E
   giving matching 1 = C   2 = A   3 = D   4 = B   5 = E

b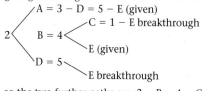

A = 3 − D = 5 − E (given)
   C = 1 − E breakthrough
B = 4
   E (given)
D = 5
   E breakthrough

so the two further paths are: 2 − B = 4 − C = 1 − E and
2 − D = 5 − E

10 a Chemical A $5x + y ⩾ 10$
   Chemical B $2x + 2y ⩾ 12$ $[x + y ⩾ 6]$
   Chemical C $\frac{1}{2}x + 2y ⩾ 6$ $[x + 4y ⩾ 12]$

b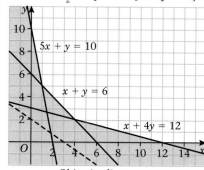

Objective line
$2x + 3y$

c $T = 2x + 3y$

d $(4, 2)$ $T = 14$

e If there were 3 or more variables the problem could not be solved graphically.
   So adding a third fertiliser Z, would mean a graphical method could not be used.

11 a

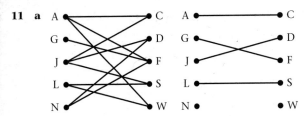

b

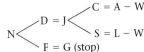

C = A − W

D = J

N

S = L − W

F = G (stop)

Breakthrough, so choose just one of these.

Change status to give

N = D − J = C − A = W giving matching

A = W   G = F   J = C   L = S   N = D

N = D − J = S − L = W giving matching

A = C   G = F   J = S   L = W   N = D

c If J does D, N must do F leaving G without a task.

12 a Maximise $P = 300x + 500y$

b Finishing $3.5x + 4y \leqslant 56 \Rightarrow 7x + 8y \leqslant 112$ (o.e.)

Packing $2x + 4y \leqslant 40 \Rightarrow x + 2y \leqslant 20$ (o.e.)

c

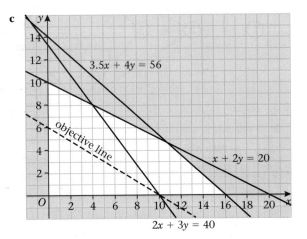

d For example, Point testing

- test all corner points in feasible region
- find profit at each and select point yielding maximum.

Profit line

- Draw profit lines.
- Select point on profit line furthest from the origin.

e Using a correct, complete method.

make 6 Oxford and 7 York profit = £5300

$(6, 7) \rightarrow 5300 \ (14.4, 1.4) \xrightarrow{\text{integer}} (14, 1) \rightarrow 4700 \ (16, 0) \rightarrow 4800$

$(0, 10) \rightarrow 5000$

f The line $3x + 4y = 49$ passes through (6, 7) so reduce *finishing* by 7 hours.

13 a

A •   • 1

B •   • 2

C •   • 3

D •   • 4

E •   • 5

F •   • 6

b There are six possible pairs of alternating paths – you must only choose one pair.

i D − 3 = A − 1 = C − 4 = B − 2 then E − 5 = F − 6

ii D − 3 = A − 1 = C − 5 = F − 6 then E − 5 = C − 4 = B − 2

iii E − 3 = A − 1 = C − 4 = B − 2 then D − 3 = E − 5 = F − 6

iv E − 3 = A − 1 = C − 5 = F − 6 then D − 3 = E − 5 = C − 4 = B − 2

v E − 5 = F − 6 then D − 3 = A − 1 = C − 4 = B − 2

vi E − 5 = F − 3 = A − 1 = C − 4 = B − 2 then D − 3 = F − 6

changing status, each of these give the same complete matching

A = 1   C = 4   E = 5

B = 2   D = 3   F = 6

Remember to update your 'initial' matching after the first pass through the algorithm. The first alternating path switches things around and your second path needs to take these changes into account.

14 a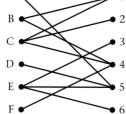

b

c B − 1 = C − 2 change status B = 1 − C = 2
giving improved matching A = 5 B = 1 C = 2 D = ? E = 6 F = 4

d For example, E is the only person who can do 3 and also the only person who can do 6, so a 1 − 1 complete matching is not possible.

15 $x = 31$ •————— highest of (11 + 8) and (17 + 14)

a $y = 17$ •————— lowest of (31 − 14), (38 − 9) and (30 − 6)

b

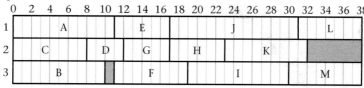

All critical activities have a zero total float.

c 107 ÷ 38 = 2.8 (1 d.p.) so at least 3 workers needed

d For example,

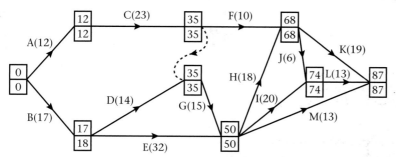

16 a

| 0 | 2 | 4 | 6 | 8 | 10 | 12 | 14 | 16 | 18 | 20 | 22 | 24 | 26 | 28 | 30 | 32 | 34 | 36 | 38 |

b A, C, E, H, J, K and L

c Total float = 35 − 17 − 14 = 4

All critical activities have a zero total float.

d *Either* 226 ÷ 87 = 2.6 (1 d.p.) so at least 3 workers needed
or 69 hours into the project activities J, K, I and M *must* be happening so at least 4 workers will be needed.

e

| 0 | 5 | 10 | 15 | 20 | 25 | 30 | 35 | 40 | 45 | 50 | 55 | 60 | 65 | 70 | 75 | 80 | 85 | 90 |

New shortest time is 89 hours.

17 a Objective: maximise $P = 30x + 40y$ (or $P = 0.3x + 0.4y$)
subject to:

$$x + y \geqslant 200$$
$$x + y \leqslant 500$$
$$x \geqslant \tfrac{20}{100}(x + y) \Rightarrow 4x \geqslant y$$
$$x \leqslant \tfrac{40}{100}(x + y) \Rightarrow 3x \leqslant 2y$$
$$x, y \geqslant 0$$

b

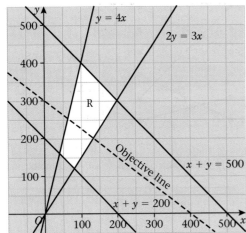

c Visible use of objective line method – objective line drawn or vertex testing – all 4 vertices tested

Vertex testing

(40, 10) → 1600

(80, 120) → 7200

(100, 400) → 19 000

(200, 300) → 18 000

Intersection of $y = 4x$ and $x + y = 500$

(100, 400) profit £190 (or 19 000 p)

18 a Critical activities are B, F, J, K and N length of critical path is 25 hours I is not critical

b Total float on A = 5 − 0 − 3 = 2

Total float on C = 9 − 0 − 6 = 3

Total float on D = 11 − 3 − 3 = 5

Total float on E = 9 − 3 − 4 = 2

Total float on G = 9 − 4 − 3 = 2

Total float on H = 16 − 7 − 7 = 2

Total float on I = 16 − 9 − 5 = 2

Total float on L = 22 − 11 − 4 = 7

Total float on M = 22 − 16 − 2 = 4

Total float on P = 25 − 18 − 3 = 4

c

1 2 3 4 5 6 7 8 9 10 11 12 13 14 15 16 17 18 19 20 21 22 23 24 25

B			F		J		K			N		
A		C										
	C	D										
		E										
		G										
			H									
			I									
			L									
				M								
				P								

d For example

Minimum number of workers is 3.

1 2 3 4 5 6 7 8 9 10 11 12 13 14 15 16 17 18 19 20 21 22 23 24 25

B		F		J		K			N	
A		E		D		I		L		P
	C		G			H		M		

19 a

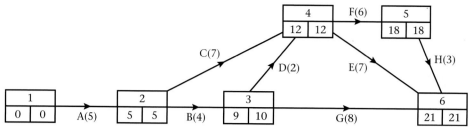

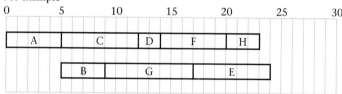

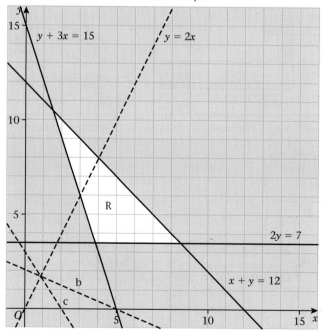

b Critical activities: A, C, F and H; length of critical path = 21

c Total float on B = 10 − 5 − 4 = 1   Total float on E = 21 − 12 − 7 = 2
Total float on D = 12 − 9 − 2 = 1   Total float on G = 21 − 9 − 8 = 4

d

e For example

Minimum time for 2 workers is 24 days.

20

b Visible use of objective line method – objective line drawn or vertex testing.
$\left[\left(3\frac{5}{6}, 3\frac{1}{2}\right) \to 25\frac{1}{6} \left(8\frac{1}{2}, 3\frac{1}{2}\right) \to 34\frac{1}{2} (4, 8) \to 48 (3, 6) \to 36\right]$
Optimal point $\left(3\frac{5}{6}, 3\frac{1}{2}\right)$ with value $25\frac{1}{6}$

c Visible use of objective line method – objective line drawn, or vertex testing – all 4 vertices tested.
$\left(3\frac{5}{6}, 3\frac{1}{2}\right)$ not an integer try $(4, 4) \to 20$ $(4, 8) \to 28$
$\left(8\frac{5}{6}, 3\frac{1}{2}\right)$ not an integer try $(8, 4) \to 32$ $(3, 6) \to 21$
Optimal point $(8, 4)$ with value 32

21 a For example, it shows dependence but it is not an activity. G depends on A and C *only* but H and I depend on A, C *and* D.

b

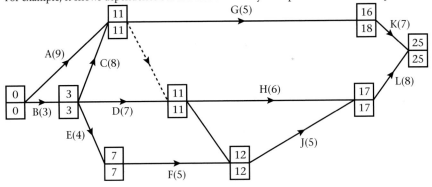

c

So B, C, E, F, I, J and L

d Total float on A = 11 − 0 − 9 = 2
Total float on D = 11 − 3 − 7 = 1
Total float on G = 18 − 11 − 5 = 2
Total float on H = 17 − 11 − 5 = 1
Total float on K = 25 − 16 − 7 = 2

e

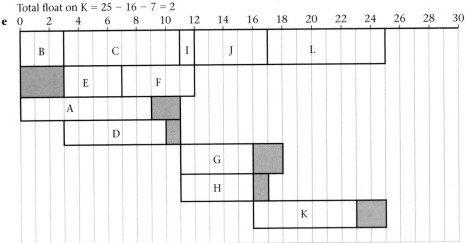

f Arithmetic lower bound = $\frac{67}{25}$ = 2.68 so a minimum of 3 workers needed.
From Gantt chart: At time 8 activities C, F, A and D must be happening, so a minimum of 4 workers are needed.
We need to take the higher of these as our best lower bound and state that a minimum of 4 workers are needed.

22 a Objective: maximise P = 0.4x + 0.2y   (P = 40x + 20y)
subject to:
$x \leqslant 6.5$
$y \leqslant 8$
$x + y \leqslant 12$
$y \leqslant 4x$
$x, y \geqslant 0$

b Visible use of objective line method – objective line drawn [e.g. from (2, 0) to (0, 4)] or all 5 points tested.
vertex testing
[(0, 0) → 0; (2, 8) → 2.4; (4, 8) → 3.2; (6.5, 5.5) → 3.7; (6.5, 0) → 2.6]
Optimal point is (6.5, 5.5) ⇒ 6500 type X and 5500 type Y

c P = 0.4(6500) + 0.2(5500) = £3700

23 a x = 12 y = 24 z = 19

b Allows J and K to be uniquely expressed in terms of their end events.

c F, E, I, J, G, H

d It would have no effect, B has a total float of 2 so a delay of one hour would still permit the project to be completed on time.

e For example,
- the total of activities is 54. 2 workers working for 24 hours would not be sufficient,
- $54 \div 24 = 2.25$, so 2 workers are not enough,
- 7 hours into the project A, B, C, D, E, F and G must be completed; these activities require 18 hours to complete them so 2 workers could not be enough.

f

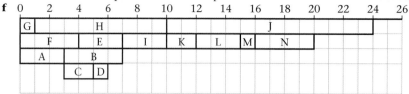

g 10 extra hours (7 + 3) so £280

24 a $x = 0$
$y = 7$ [latest out of (3 + 2) and (5 + 2)]
$z = 9$ [earliest out of (13 − 4) and (19 − 7) and (16 − 2)]

b Length is 22
Critical activities: B, D, E and L

c Total float on N = 22 − 14 − 3 = 5
Total float on H = 16 − 5 − 3 = 8

d

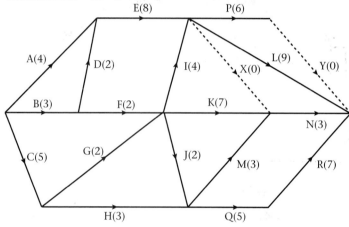

e 22 hours [critical path is still B – D – E – L]

25 a $x + y \geqslant 380$ (total production is at least 380)
$y \geqslant 125$ (at least 125 gallons of y)
$2x + 4y \leqslant 1200 \Rightarrow x + 2y \leqslant 600$ (processing time)
$x \geqslant 0$

b $C = 3x + 2y$

c

Visible use of objective line method – objective line drawn or vertex testing – all 3 vertices
[(160, 220) → 920; (255, 125) → 1015; (350, 125) → 1300]
Optimal point is (160, 220) value £920.

d (350, 125) $C = £1300$

26 a

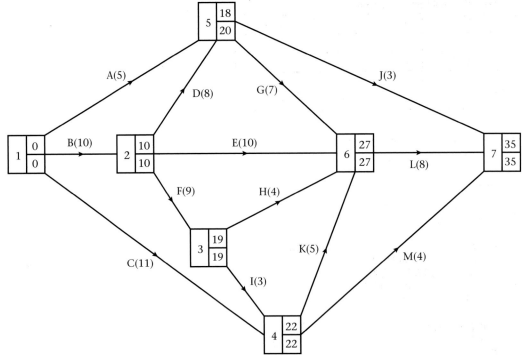

b Total float on A = 20 − 0 − 5 = 15
Total float on B = 10 − 0 − 10 = 0
Total float on C = 22 − 0 − 11 = 11
Total float on D = 20 − 10 − 8 = 2
Total float on E = 27 − 10 − 10 = 7
Total float on F = 19 − 10 − 9 = 0
Total float on G = 27 − 18 − 7 = 2

Total float on H = 27 − 19 − 4 = 4
Total float on I = 22 − 19 − 3 = 0
Total float on J = 35 − 18 − 3 = 14
Total float on K = 27 − 22 − 5 = 0
Total float on L = 35 − 27 − 8 = 0
Total float on M = 35 − 22 − 4 = 9

c Critical activities: B, F, I, K and L   length of critical path is 35 days
d New critical path is B, F, H, L   length of new critical path is 36 days

Examination style paper

1 a An alternating path starts and finishes at unmatched nodes and selects arcs that are alternately 'not in' and 'in' the initial matching

b
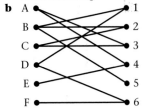

c

```
A •        • 1
B •        • 2
C •        • 3
D •        • 4
E •        • 5
F •        • 6
```

d E − 4 = A − 3 = C − 2 = B − 1 breakthrough
Change status
E = 4 − A = 3 − C = 2 − B = 1
Improved matching: A = 3   C = 2   E = 4
                   B = 1   D = 6   F unmatched

e F − 6 = D − 1 = B − 5
Change status
F = 6 − D = 1 − B = 5
Improved matching: A = 3   C = 6   E = 4
                   B = 5   D = 1   F = 6

2 a For example (Bubbling left to right)

H	L	R	A	T	J	Y	S	M
H	L	A	R	J	T	S	M	Y
H	A	L	J	R	S	M	T	Y
A	H	L	J	R	M	S	T	Y
A	H	J	L	M	R	S	T	Y

Sorted – no more exchanges

b 1st pivot $\dfrac{1+9}{2} = 5$ Mark R > M reject 1 – 5

2nd pivot $\dfrac{6+9}{2} = 8$ Tom R < T reject 8 – 9

3rd pivot $\dfrac{6+7}{2} = 7$ Sue R < S reject 7

4th pivot 6 Ramin: name found

3 a Lower bound = $\dfrac{69}{20}$ = 3.45 so 4 bins needed

b Bin 1: 8, 5, 3   Bin 3: 11   Bin 5: 12
  Bin 2: 14, 6   Bin 4: 10
c Reordering list: 14, 12, 11, 10, 8, 6, 5, 3
  Bin 1: 14, 6   Bin 3: 11, 5, 3
  Bin 2: 12, 8   Bin 4: 10

4 a $AC \left\{ \begin{matrix} CE \\ or \\ CF \end{matrix} \right\}$, EF, BE, DE

b Either   or

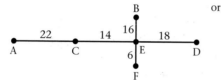   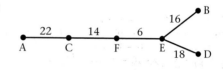

Weight: £76

5 a

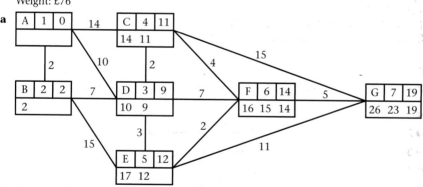

b For example
  19 − 5 = 14   FG
  14 − 2 = 12   EF
  12 − 3 = 9   DE
  9 − 7 = 2   BD
  2 − 2 = 0   AB
c (From D to F via E takes 5 km. From D to F via C takes 6 km so use this route)
  A – B – D – C – F – G length 20
6 a BC + GH = 14 + 13 = 27
  BG + CH = 14 + 12 = 36 ←
  BH + CG = 10 + 21 = 31
  Repeat BG and CH
  Route length: 157 + 26 = 183 m
b We are seeking a semi-Eulerian route. We need to repeat just one path between two odd vertices, not including the start of B.
  The options are GH − 13
  CH − 12
  CG − 21
  We chose to repeat CH so our finishing vertex is G.

7

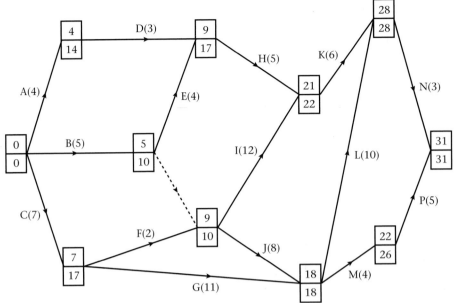

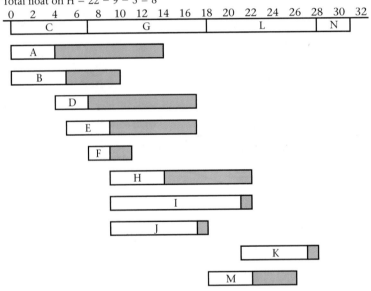

a
$$\begin{bmatrix} p = 18 & r = 28 & t = 26 & v = 22 & x = 10 \\ q = 21 & s = 22 & u = 28 & w = 18 & y = 18 \end{bmatrix}$$

b Critical activities: C, G, L, N

c Total float on E = 17 − 5 − 4 = 8
   Total float on H = 22 − 9 − 5 = 8

d

0	2	4	6	8	10	12	14	16	18	20	22	24	26	28	30	32

e On day $13\frac{1}{2}$ activities G, I and J *must* be happening.

8 a

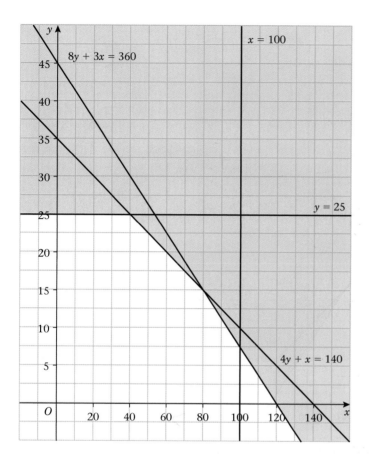

b

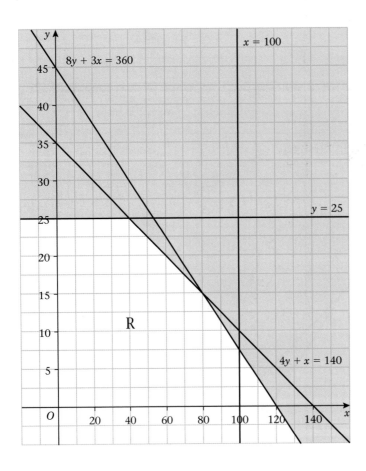

c

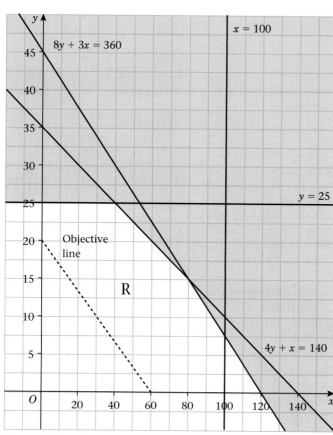

Optimal value of intersection of $8y + 3x = 360$ and $4y + x = 140$

Optimal point $(80, 15)$

d $F = 3(15) + (80) = 125$

Index

LONGLEY PARK SIXTH FORM COLLEGE
HORNINGLOW ROAD
SHEFFIELD
S5 5SG